Silicon Sensors and Circuits

Sensor Physics and Technology Series

Series editors:
Professor K T V Grattan
Centre for Measurement, Instrumentation and Applied Physics
City University
London UK

Dr A Augousti
School of Applied Physics
Kingston University
Kingston-upon-Thames, UK

The Sensor Physics and Technology Series aims to bring together in a single series the most important developments in the rapidly-changing area of sensor technology and applications. It will present a snapshot of the range of effort which is being invested internationally in the development of novel types of sensors. New workers in the area of sensor technology will also be catered for with an introduction to the subject through the provision of tutorial guides. Volumes may be sensor technology or applications oriented, and will present recent results from the cutting edge of research in a compact monograph format.

Topics covered will include:

- optical sensors: free-space sensors
- optical sensors: guided wave sensors
- solid state sensors
- biosensors
- microwave sensors
- ultrasonic sensors
- process tomography
- control of networked sensors system control and data acquisition
- medical instrumentation
- infrared sensors
- chemical and biochemical sensing
- environmental sensing
- industrial applications

Titles available

1. Biosensors
   Tran Minh Cahn

2. Fiber Optic Fluorescence Thermometry
   Grattan and Zang

Silicon Sensors and Circuits

On-chip compatibility

Edited by

R. F. Wolffenbuttel

Laboratory for Electronic Instrumentation
Department of Electrical Engineering
Delft University of Technology
The Netherlands

CHAPMAN & HALL

London · Glasgow · Weinheim · New York · Tokyo · Melbourne · Madras

Published by Chapman & Hall, 2–6 Boundary Row, London SE1 8HN, UK

Chapman & Hall, 2–6 Boundary Row, London SE1 8HN, UK

Blackie Academic & Professional, Wester Cleddens Road, Bishopbriggs, Glasgow G64 2NZ, UK

Chapman & Hall GmbH, Pappelallee 3, 69469 Weinheim, Germany

Chapman & Hall USA, 115 Fifth Avenue, New York, NY 10003, USA

Chapman & Hall Japan, ITP-Japan, Kyowa Building, 3F, 2-2-1 Hirakawacho, Chiyoda-ku, Tokyo 102, Japan

Chapman & Hall Australia, 102 Dodds Street, South Melbourne, Victoria 3205, Australia

Chapman & Hall India, R. Seshadri, 32 Second Main Road, CIT East, Madras 600 035, India

First edition 1996

Printed in Great Britain by St Edmundsbury Press, Bury St Edmunds, Suffolk

ISBN 0 412 70970 8

A catalogue record for this book is available from the British Library

Library of Congress Catalog Card Number: 95-74627

∞ Printed on acid-free text paper, manufactured in accordance with ANSI/NISO Z39.48-1992 (Permanence of Paper).

Contents

CONTRIBUTORS

J. Bryzek
Lucas/Novasensor,
1055 Mission Court,
CA 94539,
USA

P.J. French
Delft University of Technology,
Department of Electrical
Engineering,
Laboratory for Electronic
Instrumentation,
Mekelweg 4,
2628 CD Delft,
The Netherlands

H. Kaneko
Nissan Research Center,
Nissan Motor Company,
Natsushima-cho 1,
Yokosuka 237,
Japan

T. Kita
Nissan Research Center,
Nissan Motor Company,
Natsushima-cho 1,
Yokosuka 237,
Japan

T. Mihara
Nissan Research Center,
Nissan Motor Company,
Natsushima-cho 1,
Yokosuka 237,
Japan

K. Najafi
University of Michigan,
Department
of Electrical Engineering and
Computer Science,
1246 EECS, Ann Arbor,
MI 48109-2122,
USA

G. Smith
Lucas Advanced Engineering
Center,
Dog Kennel Lane, Shirley,
Solihull,
West Midlands B90 4JJ,
UK

M.J. Vellekoop
Xensor Integration B.V.
Delft University of Technology,
Department of Electrical
Engineering,
Laboratory for Electronic
Instrumentation,
Mekelweg 4,
2628 CD Delft,
The Netherlands

R.F. Wolffenbuttel
Delft University of Technology,
Department of Electrical
Engineering,
Laboratory for Electronic
Instrumentation,
Mekelweg 4,
2628 CD Delft,
The Netherlands

PREFACE

This book is the result of lectures given at the NEXUS workshop on silicon sensor realization compatible with microelectronic circuit fabrication. It is intended to provide an overview of the issues that are relevant in the pursuit of on-chip functional integration in silicon of all the functions of a data-acquisition unit. At the start of the research on the silicon sensor as a practical device many promises were made. The aspects that have been overlooked in these expectations and that have made the integrated silicon smart sensor concept appear to be the ultimate solution to all instrumentation and control problems will be identified. Disregarding all the promises of the early phase that cannot be fulfilled, the silicon sensor concept will be shown to be sound and as competitive as was perceived at the introduction of the idea. However, it is not viable in as many applications as expected and the pace of developments indicate that more effort and time is required than anticipated. The true and vain promises are distinguished and the technical feasibility, the economic viability and the limitations of the silicon smart sensor concept are identified.

In the first chapter the problem is stated; the silicon smart sensor concept is introduced, followed by a discussion of the basics of bipolar and CMOS processing in view of their suitability to be merged with sensor processing. Finally, the impact of on-chip functional integration of the silicon sensor and the readout circuits will be discussed using a selected number of popular transduction effects in silicon.

In chapter 2 the compatibility of silicon microelectronic processing and micromachining in silicon is introduced. Bulk micromachining and surface micromachining are discussed in view of their merits with respect to fabrication compatibility with already integrated active devices.

In chapter 3 ZnO is introduced as a compatible material that can be used to overcome the basic limitations of silicon as a sensor material, while maintaining compatibility with integrated circuit fabrication. ZnO makes it possible to add piezoelectric properties to silicon for the realization of piezoelectric actuators, microphones and Surface Acoustic Wave (SAW) devices.

In chapter 4 the economics of the silicon smart sensor concept are investigated. The economics governing the costs of smart sensor fabrication will be shown to be quite different from the wafer yield criterion used in integrated circuit manufacturing, due to production volume, costs of packaging, costs of bonding and the extensive testing of non-electrical parameters. It will be shown that the merit of on-chip integration depends on the added value it provides in the intended application. Special compatible sensor processes will be evaluated in view of the added value principle.

The future relation between the sensor industry and the microelectronics industry is also discussed in this chapter. As the objective for increased on-chip functional integration pushes the production volume and the required process control to the level that is already commonplace in microelectronic circuit manufacturing, the sensor industry should focus on specialization of sensor post-processing and should use the microelectronics industry as a subcontractor rather than as a competitor. Other suitable survival strategies for the sensor industry will be shown to be based on the partition of the integrated silicon smart sensor into the sensor part and the readout part until the final stage of the processing, while maintaining the physical unity of the final data-acquisition unit, using hybrid integration, wafer-to-wafer bonding or flip-chip bonding.

Chapter 5 focuses on the economics of micromechanical sensors, especially pressure sensors and accelerometers. The functional integration issue will be discussed with an emphasis on automotive applications.

In chapter 6 the medical field is discussed as the only application field in which the small size of the silicon sensor is of utmost importance. The high added value of small size and the need for integrated functions shifts the economics towards the smart sensor concept, despite the limited production volume and the complicated packaging.

In chapter 7 the merit of the smart sensor concept is evaluated from the perspective of the automotive industry. Examples of high-level functional on-chip integration are presented that have actually been implemented in large volume in the automotive, along with examples where the economics are more in favor of hybrid integration of sensor and readout circuits. Finally, in chapter 8 conclusions are drawn and recommendations are given.

Chapters 1, 4 and 8 have been written by R.F. Wolffenbuttel and the editor is indebted to all persons participating in the fabrication compatibility projects at the Laboratory for Electronic Instrumentation of the Department of Electrical Engineering of the Delft University of Technology: M. Bartek, B.P. van Drieënhuizen, J.A. Foerster, P.J. French, P.J.T. Gennissen, J.F.L. Goosen, G. de Graaf, R.P. van Kampen, T.A. Kwa, Y.X. Li, K.M. Mahmoud, D. Poenar, M.J. Rutka, P.M. Sarro (DIMES) and M.R. Wolffenbuttel. The work presented in this chapter is supported by FOM/STW.

Chapter 2 has been written by P.J. French. Thanks are due to the members of DIMES and in particular P.M. Sarro and R. Mallée for their help in developing the processes. Further thanks are due to many colleagues who have contributed greatly to the work presented here; in particular, M. Bartek, B.P. van Drieënhuizen, J.A. Foerster, J.F.L. Goosen, R.P. van Kampen, T.A. Kwa, Y.X. Li, D. Poenar, P.J. Trimp, R.F. Wolffenbuttel, M.R. Wolffenbuttel and also to D. de Haan and E.J.M. Fakkeldij of Materials Research of the Delft University of Technology.

Chapter 3 has been written by M.J. Vellekoop and the author is indebted to A. Venema for fruitful discussions about the CdS and ZnO technology and to A. Cimpoiasu for her work on the reduction of residual stress in ZnO films. Th.H. de Keyser and N.M. van der Pers of the Department of Material Science of the Delft University of Technology are gratefully acknowledged for helpful discussions and X-ray analysis of ZnO-Si layered devices.

Chapter 5 has been written by J. Bryzek and G. Smith.

The research results and material presented in chapter 6 are based on research and development of solid-state integrated sensors carried out at the University of Michigan. The participation of J. Ji, S. Cho and J. Lund is acknowledged. The support of the staff, especially T. Hull and W. Bear, of the Solid-State Electronics Laboratory is greatly appreciated. A major part of the research presented has been supervised and performed by K.D. Wise. The author also acknowledges the many contributions of D.J. Anderson, J. Hetke, S.L. BeMent of the Biomedical Laboratory of the University of Michigan. This work has been funded by the Neural Prosthesis Program, the National Institute for Neurological Disorders and Stroke and the National Institutes of Health. The author is grateful for the continuing support and encouragement of T.F. Hambrecht and W.J. Heetderks of NIH.

Chapter 7 has been written by P.J. French, H. Kaneko, T. Kita and T. Mihara.

The editor is indebted to M.J. Rutka and M.R. Wolffenbuttel for their assistance in manuscript preparation.

R.F. Wolffenbuttel
Delft University of Technology
April 1995

1
ON-CHIP FUNCTIONAL INTEGRATION

1.1 Introduction

Research on silicon sensors started about 25 years ago as a kind of spin-off of mainstream research on silicon microelectronic technology and circuits [1.1]. In the early phase the device was not identified as a silicon sensor, but rather as a curiosity or at best a means for testing semiconductor properties. Nowadays the silicon sensor field has gained general acceptance and has become a mature multi-disciplinary science. Two basic incentives have been decisive in this development, viz. the prospect of low unit-price of the sensor in mass production, similar to the economies of scale in microelectronic circuits, and the anticipated feasibility of realizing the complete data-acquisition system unit on the smallest possible material carrier; the microelectronic silicon chip. Naturally, the researchers at universities have, in their creative innocence, been mainly lured by their scientific curiosity on the material properties of silicon and were anxious to take advantage of the unique opportunities offered by the mere presence of a huge silicon integrated circuit manufacturing industry, both in terms of availability of high-purity substrate and the infrastructure of suppliers of high-performance equipment. However, the appealing commercial prospects mentioned have been unscrupulously (ab)used, manipulated, refined and rephrased in the numerous proposals for grants for funding of silicon sensor research that have passed the offices of the various national funding agencies in many countries during these years. When the experienced applicant also discusses the indispensability of a measurement unit in applications of great public concern, such as safety of consumer products or protection of the environment, it becomes extremely difficult to reject the proposal.

Confronting a silicon sensor researcher with THE existential question related to the profession is, therefore, most unlikely to lead to an embarrassing silence, as the person involved has to be aware of the one-and-only dogma in the field to avoid confrontation with any existential hazard of a more personal nature. Unfortunately, funding agencies have become slightly less receptive over time to these arguments, possibly due to the limited proven success of this on-chip integrated silicon smart sensor concept, and have started questioning its validity.

In terms of academic infrastructure the research on silicon sensors could be regarded as mature. Journals have been established to present the results on sensor research. Moreover specialized journals are available on sub-topics such as chemical sensors and micromechanics. National and international conferences on silicon sensors would fill the calendar. However, if the silicon sensor concept was as sound as it has been claimed for two decades, it should also have given rise to a considerable industrial infrastructure by now. This promise has, unfortunately, not been fulfilled. In many applications the conventional sensor remains the *de-facto* industrial standard and is far from being replaced by its sophisticated silicon counterpart. Many silicon sensor foundries have been enthusiastically started; however, only a few small and medium sized companies have managed to find their niche. This present situation has also given rise to criticism of the validity of the silicon smart sensor concept.

This book is about the real prospects of the silicon sensor and its celebrated compatibility with microelectronic circuits in silicon. It will be shown that the foundations of the silicon smart sensor concept are sound and it will be demonstrated that such a device is both technologically feasible and economically viable. Issues to be addressed are not only the advantages of silicon sensors for the realization of on-chip functional integration of a complete data-acquisition unit in silicon, which have proven to be genuine, but also the constraints that were overlooked in the beginning. It will be demonstrated that the material compatibility is a necessary, but insufficient, prerequisite for the merging of sensor and readout circuits and that the essential issue is rather their fabrication compatibility. Furthermore, the wafer yield criterion, which is widely used for estimating the economic viability of integrated circuits, will be demonstrated to be an inadequate tool for silicon sensors because of the differences in production volume, packaging and testing.

The dilemma of the silicon sensor industry is also discussed. Small and medium-sized sensor foundries are well equipped to perform a limited number of special processing steps, such as micromachining, but do not have a microelectronic production line operational. Utilizing the compatibility to actually merge functions on-chip, to exploit the silicon advantage compared to conventional sensors in order to gain market acceptance, imposes increasing requirements on process control. At exactly the stage where the market becomes interested in the smart sensor concept, the production volume becomes interesting for a major integrated circuit foundry. The equipment and process control required for smart sensor production is already commonplace in integrated circuit foundries, whereas the sensor industry would have to evolve to that situation. Obviously, a free market economy will not grant the sensor industry the necessary time to develop. This phenomenon puts the silicon sensor industry in a somewhat uncomfortable position in the food chain.

It will be shown that most of the arguments initially claimed in favor of silicon sensors are in principle correct, but do not always apply due to unforeseen constraints. The major error of judgement will be demonstrated to be the adoption of the economies of scale from the microelectronics industry. Furthermore, the functional integration of all the parts of a data-acquisition unit is technologically feasible and economically viable; however, not in as many applications as anticipated and after a much longer development period than expected. Sensor processing should be designed as highly specialized pre- or post-processing steps on a standard microelectronic process. It will be shown that the mass-production capability of the major circuit foundries with

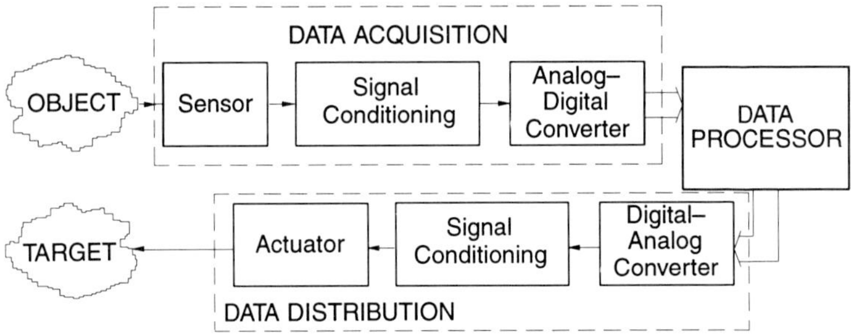

Figure 1.1. *Structure of a measurement and control system.*

their superior equipment is in that case not a threat to a sensor industry that has found its niche, but rather another convenient addition to an infrastructure that is traditionally exploited by the sensor community, as it enables a small sensor company to use the microelectronics industry as a subcontractor.

1.2 The smart sensor concept

A sensor is part of a data-acquisition system, which is in turn part of an electronic measurement and control system as shown in Fig. 1.1. The data-acquisition system is designed to acquire information from an object (phenomenon, process or system of physical, chemical or biological nature) and to transform this data into a format that is suitable for manipulation in the data processor [1.2]. The desired information is contained in an energetic signal domain. In the case of a non-electrical signal carrier (mechanical, radiant, magnetic, chemical or thermal), a sensor is required, which, based on some transduction effect in the material used, enables the signal to change from energetic carrier, without loss of information content of the intended signal source or interference from any undesired signal source. These conditions are, of course, only met up to a certain extent in a practical sensor, which exhibits offset, non-linearity, cross-sensitivity to parasitic non-electrical signals sources at the input and a susceptibility to electrical noise at the output terminals. Signal conditioning circuits are implemented to reduce these effects by offset cancelation, non-linear inverse transfer functions, compensation using multi-sensor information and output impedance transformation, amplification or filtering, respectively.

The data-acquisition structure shown in Fig. 1.1 is not the only possible approach. Integrated smart sensors, e.g. based on bitstreams using Delta-Sigma modulators, have been investigated for providing a digital output signal [1.3]. A microprocessor is typically used for data processing, so Analog-to-Digital (A/D) conversion is required. Obviously, next to reducing the effect of sensor non-idealities, the amplification in the signal processing unit should be designed to match the sensor output signal dynamic range to that of the A/D converter input. Therefore, the analog functions to be implemented in the signal processing part are highly sensor-specific, which brings up the idea of realizing the complete data-acquisition system on the same material carrier by merging the sensing element and the microelectronic readout circuits on one silicon chip. Such a device is often referred to as an 'integrated silicon smart sensor'.

Silicon microelectronic technology has evolved in the last decades towards a means for high-density integration of complex analog and digital functions. Large digital circuits have been integrated in silicon using advanced MOS processes. Demanding analog specifications are met by fabrication in a bipolar process. Digital and analog integrated circuits originating from these technologies have experienced an unprecedented market acceptance. In retrospect it can be concluded that the rapid penetration of these products in electronic systems is due to the fact that such an integrated circuit can be regarded as a building block of a high level of abstraction with respect to the details of the internal circuits. A specification of a digital integrated circuit in terms of logic levels on the in- and output terminals, combined with a functional description of input/output sequences including practical timing constraints, along with the range of power supplies, is sufficient for implementation and has strongly contributed to this rapid acceptance. Although to a smaller extent, a similar mechanism applies for analog integrated circuits. A specification in terms of transfer function and characteristic input and output parameters, such as offset, impedance, equivalent input noise sources, voltage range, common-mode performance and frequency range, is sufficient for application, without the user having to go into the details of the internal operation.

This approach can basically also be applied to the silicon sensor. One of the major disadvantages of state-of-the-art silicon sensors is that, like non-silicon sensors, only the basic sensing element is available on the material carrier and the designer or user of a measurement and control system has to go into the details of the sensor non-ideality to add the remaining parts of the data-acquisition system. The electrical output can be a voltage, a change in capacitance, etc. and is strongly dependent on the type of sensor. Moreover, only the output signal of the basic sensor element is available, which implies that no compensation of the undesired characteristics mentioned is implemented.

The integrated silicon smart sensor concept enables the on-chip integration of circuits that provides a sensor signal not corrupted by non-idealities of the basic sensing element. Moreover, self-calibration and self-test could be implemented and the output signal would comply with a standard serial bus protocol.

The improvement of user acceptance pushes for the case integration of sensor and readout circuits in one indivisible package. The logical consequence of this marketing advantage is not necessarily on-chip integration, as a separate silicon sensor chip and microelectronic chip for readout can also be placed in one package (hybrid integration). In the next section on-chip integration is compared with hybrid integration.

1.3 Merits and limitations of integrated silicon sensors

The silicon sensor in general, and the on-chip integrated silicon smart sensor concept introduced in the previous section in particular, have been promoted as having five advantages, viz.:

1. The material compatibility that enables on-chip functional integration with microelectronic circuits in silicon.
2. The low unit-costs in mass production.
3. The small size.

4. The material properties and the many transduction effects available.
5. The silicon microelectronic infrastructure available.

These claimed advantages are evaluated in more detail in this section in order to judge whether these are genuine or not.

The material compatibility with silicon microelectronic circuits is indeed a unique advantage of silicon sensors and suggests a smooth merging of the two parts. As mentioned in the previous section, this property in principle enables the fabrication of complete data-acquisition systems on a single silicon chip. However, this material compatibility of silicon sensors with microelectronic circuits in silicon is a necessary but insufficient prerequisite for such an integration. The essential issue is rather the fabrication compatibility of sensor, sensor-related analog microelectronic circuits and digital interface circuits. Fabrication compatibility includes merging of sensor processing with analog and digital microelectronic circuit processing, both on the wafer level, in terms of doping profile, thermal budget and equipment utilization, and on the die level, with respect to testing, dicing, bonding and packaging.

The main problem in such an objective is to actually merge the processes that are run to fabricate the various components. Adding one type of device should not interfere with proper operation of the other. This problem was already relevant in the early 1980's, when mixed analog/digital designs were to be merged in order to reduce the number of chips in electronic signal processing systems, such as those used in telecommunication. The active components in a MOS process are optimized for switching applications and demonstrate a poor performance when used for linear amplification, whereas bipolar transistors are very suitable for the realization of analog functions, but less equipped for the high-density integration of digital functions. This fabrication compatibility issue — the merging of analog and digital circuitry — has resulted in the BiCMOS process, which combines the merits of bipolar processing for analog functions with the high-density integration of digital functions in a CMOS process, without loss of either performance and at the expense of a slight increase in process complexity.

An integrated silicon smart sensor imposes even higher demands on the compatibility of the processes involved. Not only are the analog and digital circuits to be merged, but also the sensor has to be incorporated. The degree to which the fabrication of smart sensors is hindered by the compatibility problem is strongly dependent on the actual non-electrical signal domain. Many types of silicon sensors require a processing for maximum sensor performance that is not compatible with the fabrication of active devices of maximum performance. Frequently, the vertical doping profile required for the sensor is not suitable for that of a merged transistor. As an example, a photodiode can, in principle, be fabricated in a process using almost any junction; however, an enhanced short-wavelength response requires a very shallow junction, which cannot be used as an active layer in a bipolar transistor.

Another compatibility constraint that has often been overlooked involves equipment utilization. Wet-etching of a wafer in an alkaline solution such as KOH is usually not permitted in a standard circuit fabrication line. In general, procedures have been established that prevent the use of such etchants in the cleanroom or the wafers being re-entered, even after cleaning, for further processing subsequent to KOH etching. As a consequence the bulk micromachining has to be performed as a post-processing step [1.4]. Pre- or post-processing will be shown to be a promising strategy in general. Clearly, the material compatibility is not as trivial as initially assumed and this advantage should be considered with some caution.

Nevertheless, for the silicon sensor to gain market acceptance this should either offer superior performance compared to the conventionally applied sensor or offer the advantage of user friendliness due to the high level of abstraction of a sensor element by on-chip system integration, so that it can be regarded as a building block. The trend in silicon sensor research has to be directed at finally exploiting the potential of the material compatibility and to merge the silicon sensing element with electronic circuits in silicon on one chip to realize a smart sensor with a standard digital microprocessor-compatible output on which the undesirable characteristics of the sensing element do not appear.

The most striking advantage often claimed is the prospect of low unit-price in mass production. A tested and encapsulated integrated circuit in a standard process in silicon can be manufactured for only US\$ 0.10 mm^{-2}. Even if a silicon sensor could be designed to be fully compatible with a standard process (without adding extra layers or sensor processing on the wafer level), so that it can be easily merged, such a price is very unlikely ever to be achieved for a smart sensor, because of the production volume, special packaging and extensive electrical and non-electrical testing. The smart sensor is an on-chip data-acquisition unit that is part of a measurement and control unit. The smart sensor is to be supplied to the instrumentation and measurement industry, which is highly diverse. The volume is, therefore, not large enough, so that the economies of scale that govern the economics of integrated circuits in silicon do not apply. Moreover, testing of integrated circuits is mainly electrical. Testing of non-electrical environmental parameters, such as temperature and humidity, is mainly performed to set the operating range. The silicon sensor, however, is an intermediate between a non-electrical and an electrical parameter and extensive testing in both signal domains is required to characterize the device. Finally, the packaging of an integrated circuit solely serves to protect the chip against environmental hazards, whereas the sensor is by definition required to have some kind of interaction with the environment. The sensor package is, therefore, more application specific and due to the resulting small volume, more expensive than the standardized molded plastic encapsulation used for integrated circuits.

Therefore, wafer yield is not the dominating fabrication cost in sensor production. It is important to conclude that the on-chip data-acquisition unit is simply not as universal a building block as the logic gate or the operational amplifier. This aspect is discussed in more detail in chapter 4.

The third often claimed advantage of the silicon sensor is its small dimensions. Indeed the batch fabrication allows sensor dimensions of the order of μm. However, the small size is in many applications considered as a problem that comes with the batch processing and makes the unpackaged device difficult to handle. The two exceptions are in infrared detectors and biomedical applications. In an infrared detector the dimensions of the detector determine the thermal time constant and thus the response time. Moreover, in a thermal imager small elements are thermally isolated using micromachining and placed in a matrix to result in an

IN \ OUT	rad.	mech.	therm.	elect.	magn.	chem.
rad.	photo-lumin	radiant pressure	radiant heating	photo-electricity	photo-magneticity	photo-chem.
mech.	photo-elasticity	spring deflection	friction heating	piezo-electricity	magneto-strict.	pressure ind. explosion
therm.	incan-descence	thermal expansion	heat conductivity	Seebeck effect	Curie-Weiss law	endothermal reaction
elect.	electro-lumin	piezo-electricity	Peltier effect	pn-junction effect	Ampere's law	electrolysis
magn.	Faraday effect	magneto-strict.	Ettinghausen effect	Hall effect	magnetic induction	
chem.	chemo-lumin	explosion reaction	exothermal reaction	Volta effect		chemical reaction

Figure 1.2. *Six by six matrix showing transduction effects.*

array that requires on-chip multiplexing. In biomedical applications such as implants and in catheters, where very little space is available, sensor dimensions are of utmost importance. This field also potentially benefits the most from the smart sensor concept, e.g. in applications such as multichannel neural recording probes. However, it also imposes the most severe requirements on the sensor package and is, therefore, discussed in more detail in chapter 6.

The fourth advantage of silicon as a sensor material lies in the transduction effects that are available to realize sensors. Figure 1.2. gives an overview and shows that transduction effects from the non-electrical to the electrical domain (sensing effects) are available for all signal domains [1.5]. Silicon exhibits significant photoelectric and thermoelectric effects. Moreover, it has a reasonable carrier mobility ($\mu_n = 1.5\ m^2sV^{-1}$) that enables a significant Hall effect. Nevertheless, for all the transduction effects found in silicon, another material could be found that demonstrates that particular effect in a more pronounced way. In general silicon can be referred to as an outstanding allrounder, that is, however, always beaten by at least one contender in each speciality. Unfortunately, the sensor market does not appreciate the fact that the more successful contender is a different material in each comparison, but rather the best material is chosen in each particular application.

Moreover, silicon is non-piezoelectric, non-magnetizable, unsuitable for efficient photon emission due to its indirect bandgap and only suitable for operation up to 200°C due to the magnitude of the bandgap. The many transduction effects in silicon are, therefore, an insufficient advantage to ensure success. Obviously, the mere existence of these effects is an essential prerequisite to consider functional on-chip integration, which justifies the present extensive research activities on silicon sensor elements. However, as this basic sensor does not exhibit superior transduction performance compared to the conventional type, the compatibility aspect has to be included for a silicon sensor to become a competitive device. Again, the fabrication compatibility of silicon sensors and microelectronic circuits in silicon is the key issue and in the final section of this chapter an outline of promising transduction effects is given with an emphasis on their suitability for on-chip functional integration.

The only highly ranked material properties of silicon are the mechanical specifications. Not only does it offer a high modulus of elasticity ($E = 15 \cdot 10^{10} Nm^{-2}$), but also a high thermal stability and low hysteresis. This makes the realization of high-performance micromechanical structures possible using the micromachining techniques to be discussed in chapter 2. As will be shown in chapter 5, this has led to commercially available silicon pressure and acceleration sensors, but not to smart devices.

The silicon infrastructure resulting from the huge microelectronics industry mentioned as the fifth advantage is inarguably of enormous importance to silicon sensor research. Firstly the material is available at a purity and cost that is not possible in sensor fields based on a different material. Wafers of a well-controlled impurity concentration (10^{12}–$10^{16}\ cm^{-3}$) and a very limited number of defects ($< 100\ cm^{-2}$ [1.6]) are routinely fabricated for the microelectronic industry and are also commercially available for the sensor industry and the research groups on silicon sensors. Moreover, the sophisticated equipment required for integrated circuit fabrication has resulted in an infrastructure of suppliers, which is only economically viable due to the size of the microelectronics industry and its relatively high innovation rate.

The ever-decreasing feature size and the increasing density of active devices on a chip pushes developments, e.g. in photolithographic equipment. The silicon sensor industry benefits from this situation in two ways. Firstly, the size of the sensor industry and sensor research in terms of the available budget for equipment purchase would be inadequate to build up and maintain the necessary infrastructure of equipment suppliers. Secondly, due to the less severe requirements imposed in a silicon sensor on minimum feature size compared to an integrated circuit and the relatively large innovation push in the microelectronics industry, which causes the most advanced equipment to have that status for only a relatively short period of time, the required investment in the sensor industry is relatively moderate.

Although some small modifications have had to be made over the years on the initially claimed advantages of silicon sensors, silicon is indeed a material with a huge potential. In order to make the promise true at last, the compatibility issue has to be taken seriously and the work has to be concentrated on application areas in which the advantages of silicon really matter. This implies cashing-in the material compatibility of silicon sensors with microelectronic circuits in silicon, which implies ensuring fabrication compatibility, even if such would be at the expense of performance of the basic sensor element. The impact of the fabrication compatibility problem imposed by sensors in the various signal domains, when using standard bipolar or CMOS processing, are discussed in the next section. Focusing on applications in which the smart sensor concept gives a maximum added value would favor its implementation, e.g. in capacitive sensors in which the length of connection wires has to be minimized in order to limit the stray capacitance, in matrix sensors in which integrated addressing and multiplexing circuits limit the number of bonding pads, or in biomedical applications. Chapters 6 and 7 discuss in more detail these most promising application areas.

1.4 Standard bipolar and CMOS processing

The basics of simplified bipolar and CMOS microelectronic processing are discussed in this section. The most sensitive parts of the active structures with respect to sensor-related post-processing steps, such as micromachining and stress annealing, are identified. A standard process is referred to as being compliant when the parameters of the active devices realized therein are not very sensitive to additional sensor processing. Usually, a multi-purpose process with conservative design rules is more compliant than a high-performance special standard process.

1.4.1 Bipolar processing

The typical 2-dimensional doping profile of the two most important bipolar components, the vertical npn and the lateral pnp, is shown in Fig.1.3. Fabrication starts with a p-type substrate on which oxide is grown and patterned to define the areas where arsenic is to be driven into the substrate to form the buried-n layer [1.7]. This layer is included to minimize collector series resistance in the vertical npn and to minimize the common-base current gain, α_F, of the parasitic vertical pnp formed by the collector and base of the lateral pnp with the substrate. Subsequently, the oxide is stripped and an n-epitaxial layer of typically 4–6 μm thickness is grown to result in a structure as shown in Fig. 1.4a. Note that the buried layer diffuses into the epilayer. This effect is referred to as autodoping and limits the minimum thickness of the epilayer and the maximum processing temperature after completion of the epitaxial growth. The device separation takes place using either oxide isolation based on LOCOS, trench isolation or using junction isolation based on deep boron diffusion [1.8]. Although trench isolation is often used in advanced bipolar processing for minimizing parasitic capacitances in order to enhance high-frequency performance, junction isolation is commonly used for integrated smart sensors. After the device isolation the deep-n collector contact is defined, followed by the intrinsic and extrinsic (npn) base implant, which results in a typical doping profile as depicted in Fig. 1.4b. The extrinsic base layer is relatively thick and highly doped to minimize base series resistance, whereas the intrinsic base layer (the part underneath the emitter) is

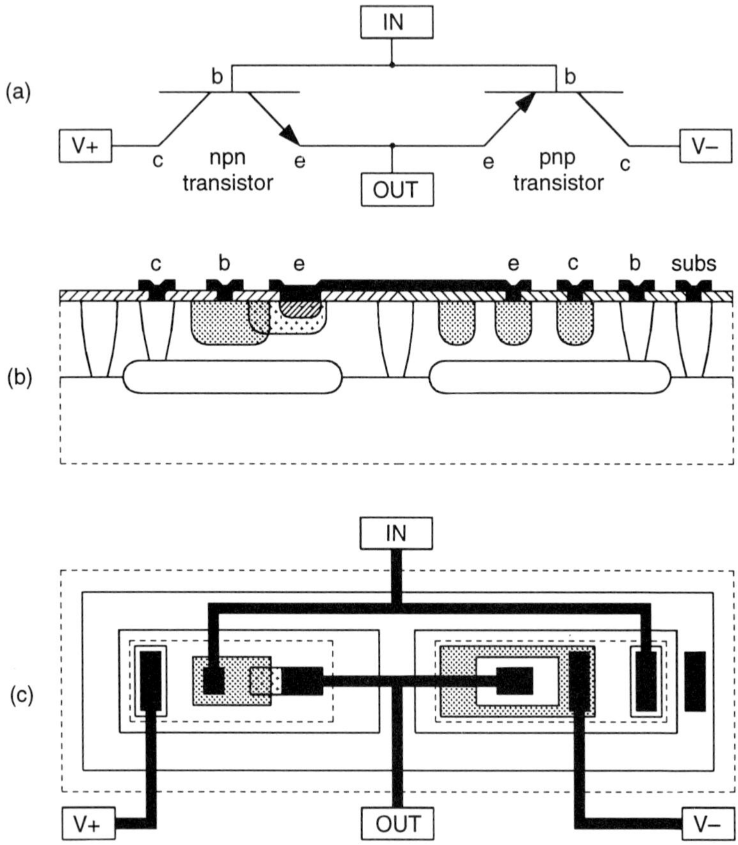

Figure 1.3. *Simple integrated bipolar buffer.*

lighter doped to result in a large common-emitter current gain, β_F. Finally, the emitter implantation is carried out, the contact holes are defined and the metal interconnect is deposited and patterned to result in the completed integrated bipolar microelectronic circuit depicted in Fig. 1.4c.

A fundamental issue that should be addressed here is the definition of the 'standard process'. In principle any departure from

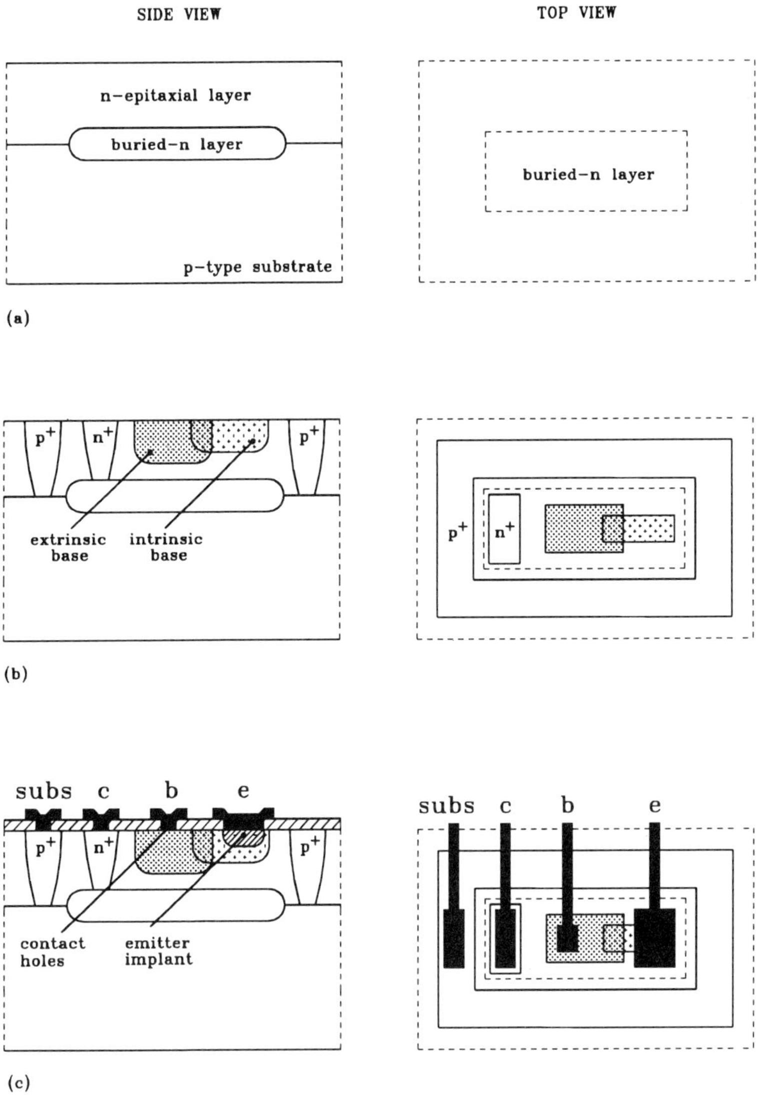

Figure 1.4. *Bipolar process flow.*

a minutely prescribed sequence of procedures causes the process to become non-standard. Although fundamentally correct, strict adherence to this view leads to large series of similar processes each devoted to a special application and thus does not offer any opportunity for the design of a more generally applicable and sensor processing-tolerant microelectronic process. Therefore, the term 'compliance of a process' is introduced here to indicate the sensitivity of the key specifications of a microelectronic process to sensor processing-induced modifications. Obviously, the more drastic the modification, the less likely it is that it can be accommodated by the standard process.

The sensor processing is usually implemented after completion of the contact hole etching, although bulk micromachining, for example, is often possible after completion of the entire microelectronic processing (chapter 2). Note that this approach implies that it is not quite like post-processing, as the metal interconnect has not yet been deposited. This seems like a rather trivial constraint considering the complexity of the metal deposition and patterning compared with the preceding processing. However, this requirement can be the cause of serious complications in the case where a sensor company subcontracts a major microelectronics foundry to fabricate the wafers containing the readout circuits, which are afterwards to be submitted to in-house sensor processing. The microelectronics manufacturer assumes responsibility only for the specifications of the completely processed devices (including interconnect), whereas stripping of the metal before the sensor processing leaves surface contaminants that are often unacceptable in many types of sensor processing.

The more fundamental fabrication compatibility problems caused by sensor processing subsequent to completed microelectronic fabrication are: (a) the increased thermal budget of already integrated active devices due to sensor processing steps such as polysilicon stress anneal or wafer-to-wafer fusion bonding; (b) modifications in the epilayer doping concentration and/or thickness that are required to enable sensor operation.

These modifications influence transistor specifications such as: (1) the common-emitter current gain, $\beta_F = \partial i_c / \partial i_b$, where i_c and i_b denote the collector and base AC current respectively; (2) the Early voltage, $V_A = -I_c/(\partial i_c / \partial v_{cb})$, (3) the collector-base breakdown

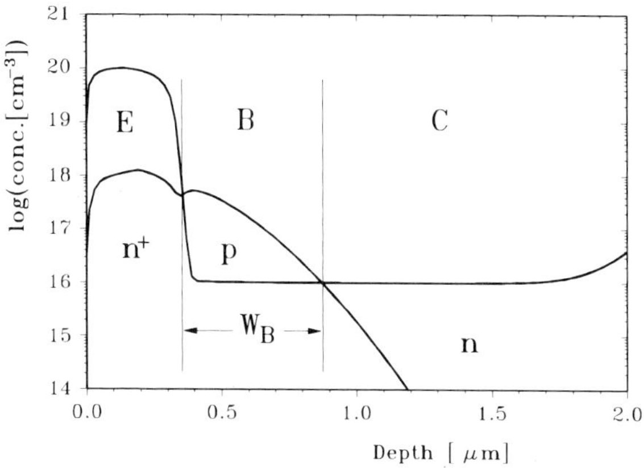

Figure 1.5. *Typical vertical doping profile of an npn bipolar transistor.*

voltage, $V_{\mathrm{cb,br}}$ and the knee current ($I_{\mathrm{c,k}}$ above which β_{F} reduces with I_{c}) [1.9].

The technology parameters, such as doping levels, temperature and epitaxial layer thickness, are optimized in a standard bipolar process to give high-performance active components. Since the vertical npn transistor is the most important element, the process parameters are designed mainly to yield a high-performance npn transistor with a reasonable performance of other active components, such as the lateral pnp and the JFET transistors. Figure 1.5 shows a typical doping profile of an integrated npn bipolar junction transistor, where the basewidth, W_{B}, is defined as the distance between the two junctions. From the figure it can be seen that W_{B} depends largely on the epitaxial layer doping concentration, provided that the other process parameters are kept constant.

The common-emitter forward current gain, β_{F}, can be expressed in terms of technology parameters as:

$$\beta_{\mathrm{F}} = \left(\frac{W_{\mathrm{B}}^2}{2L_{\mathrm{B}}^2} + \frac{D_{\mathrm{E}} W_{\mathrm{B}} N_{\mathrm{B}}}{D_{\mathrm{B}} L_{\mathrm{E}} N_{\mathrm{E}}}\right)^{-1} \rightarrow \beta_{\mathrm{F,npn}} = \left(\frac{W_{\mathrm{B}}^2}{2L_{\mathrm{n}}^2} + \frac{D_{\mathrm{p}} W_{\mathrm{B}} N_{\mathrm{B}}}{D_{\mathrm{n}} L_{\mathrm{p}} N_{\mathrm{E}}}\right)^{-1}, \tag{1.1}$$

where $L_{\mathrm{B}} = L_{\mathrm{n}}$ and $L_{\mathrm{E}} = L_{\mathrm{p}}$ denote the diffusion length of minority charge carriers in the base and the emitter, respectively,

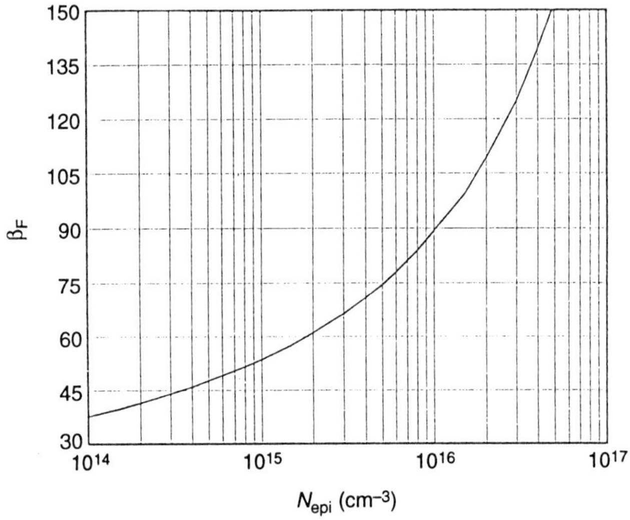

Figure 1.6. *β_F of the npn transistor vs.* N_{epi}.

and D_n and D_p are the diffusion coefficients, W_B the base width and N_B the base doping concentration. Practical values for the npn transistor are: $L_n \simeq 4\mu m$, $L_p \simeq 1\mu m$ [1.10] and $D_n/D_p = \mu_n/\mu_p = 1100/400$ [1.11]. Sensor operation might require modifications in the epilayer doping concentration. The common-emitter current gain as a function of epitaxial doping concentration is shown in Fig. 1.6.

The epitaxial layer serves as the collector. As the base doping profile remains unaltered, the basewidth increases with decreasing epilayer concentration. The width of the epilayer has no effect on β_F as long as the buried does not touch the base layer.

The common-emitter current gain of the lateral pnp can be derived using a similar expression (the epilayer forms the base in the lateral pnp and the extrinsic baselayer of the vertical npn is used for both the collector and the emitter in the lateral pnp, $N_{E,pnp} = N_{B,npn}$). Therefore,

$$\beta_{F,pnp} = \left(\frac{W_B^2}{2L_p^2} + \frac{D_n W_B N_{epi}}{D_p L_n N_B} \right)^{-1}, \tag{1.2}$$

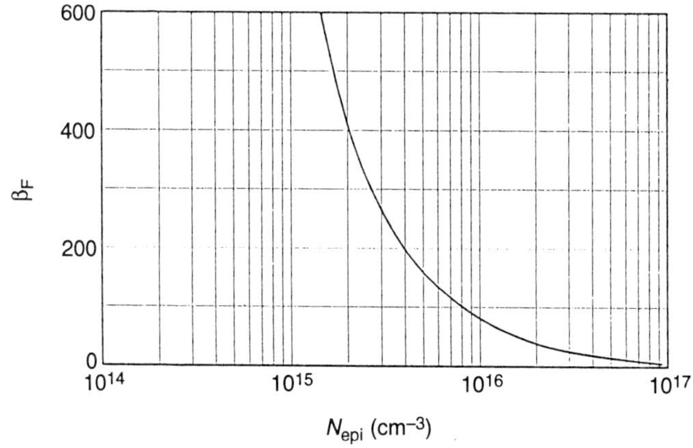

Figure 1.7. β_F *of the lateral pnp transistor vs.* N_{epi}.

where N_{epi} denotes the epilayer doping concentration.

The basewidth, W_B, of the lateral pnp transistor is equal to the emitter-to-collector mask separation minus the lateral diffusion of the emitter and collector layer. The diffusion of minority charge carriers in the epilayer, $L_p \simeq 10\mu m$. Furthermore, the doping dependence of the mobilities has to be taken into consideration, $D_n/D_p = \mu_n/\mu_p = 400/470$. The common-emitter current gain of the lateral pnp is shown in Fig. 1.7 as a function of the epitaxial layer doping concentration. The collector and emitter lateral and vertical doping profiles remain unchanged, which implies that the junction moves into the base at a decreasing epilayer concentration. Therefore, the base becomes narrower with decreasing epilayer concentration, resulting in an increasing base transport efficiency and thus an increasing β_F. Secondly, the emitter injection efficiency increases due to the increasing ratio between emitter and base doping concentration, N_E/N_B; however, the collector collection efficiency decreases for the same reason. As the pnp transistor action takes place laterally close to the silicon surface, the thickness of the epilayer does not affect β_F.

The next bipolar transistor specification that is influenced by sensor post-processing is the breakdown voltage. The breakdown

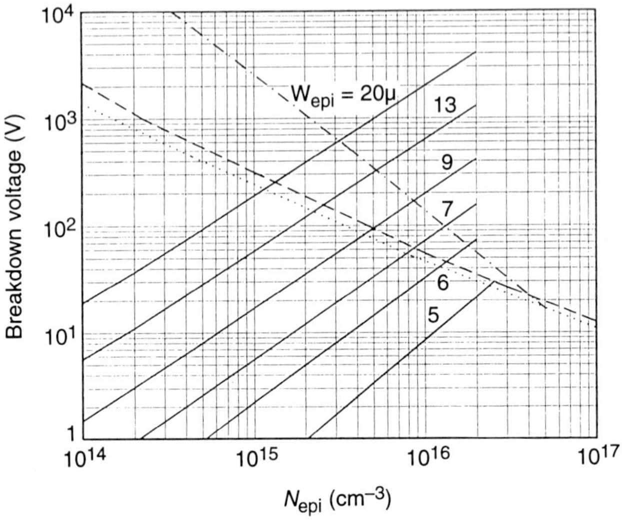

Figure 1.8. *Breakdown voltage of the npn transistor vs.* N_{epi}.

voltage of a single-sided abrupt junction due to avalanche multiplication is given by:

$$V_{cb,br} = \frac{\epsilon_{Si} {E_{crit}}^2}{2qN_{epi}}, \tag{1.3}$$

where E_{crit} denotes the critical electric field at breakdown, q the elementary charge and ϵ_{Si} the permittivity of silicon. The breakdown voltage, therefore, decreases with epilayer doping concentration. However, breakdown can also occur due to full depletion of the epilayer. Obviously, full depletion is achieved at a larger reverse voltage in the case of an epilayer of larger thickness or larger doping concentration. Figure 1.8 shows the plot of the breakdown voltage as a function of the epilayer doping concentration for several mechanisms. The dashed curve indicates breakdown due to avalanche multiplication. The solid lines show breakdown due to full depletion of the epilayer and the chain line shows breakdown due to punch-through.

In the lateral pnp, the base is the lighter-doped layer and break-

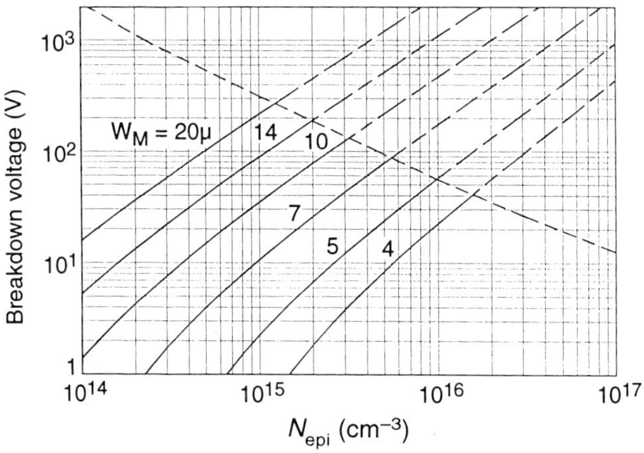

Figure 1.9. *Breakdown voltage of the lateral pnp transistor vs.* N_{epi}.

down is primarily due to punch-through. Figure 1.9 shows the breakdown voltage of the lateral pnp as a function of the epitaxial layer doping concentration. The dashed curve indicates breakdown due to avalanche multiplication and the solid curves show breakdown due to punch-through between emitter and collector with the collector-to-emitter mask separation, W_{M}, as a parameter.

The Early voltage is defined as:

$$V_{\mathrm{A}} = -\frac{I_{\mathrm{c}}}{\partial I_{\mathrm{c}}/\partial V_{\mathrm{cb}}} = \frac{Q_{\mathrm{B}}}{C_{\mathrm{cb}}}, \tag{1.4}$$

where Q_{B} denotes the majority carrier charge in the base per unit area and C_{cb} the collector-base depletion capacitance. Therefore:

$$\begin{aligned} V_{\mathrm{A,npn}} &= \frac{Q_{\mathrm{B}}}{C_{\mathrm{cb}}} = \frac{qN_{\mathrm{B}}W_{\mathrm{B}}}{\sqrt{\frac{q\epsilon_{\mathrm{Si}}N_{\mathrm{B}}N_{\mathrm{epi}}}{N_{\mathrm{B}}+N_{\mathrm{epi}}}\frac{1}{2\varphi_{\mathrm{f,cb}}}}} \\ &= \sqrt{\frac{2qN_{\mathrm{B}}{W_{\mathrm{B}}}^2(N_{\mathrm{B}}+N_{\mathrm{epi}})}{\epsilon_{\mathrm{Si}}N_{\mathrm{epi}}}\varphi_{\mathrm{f,cb}}} \quad , \end{aligned} \tag{1.5}$$

where $\varphi_{\mathrm{f,cb}}$ denotes the build-in potential, $\varphi_{\mathrm{f,cb}} = (kT/q) \times \ln(N_{\mathrm{B}}N_{\mathrm{epi}}/n_{\mathrm{i}}^2)$. The Early voltage of the vertical npn transistor

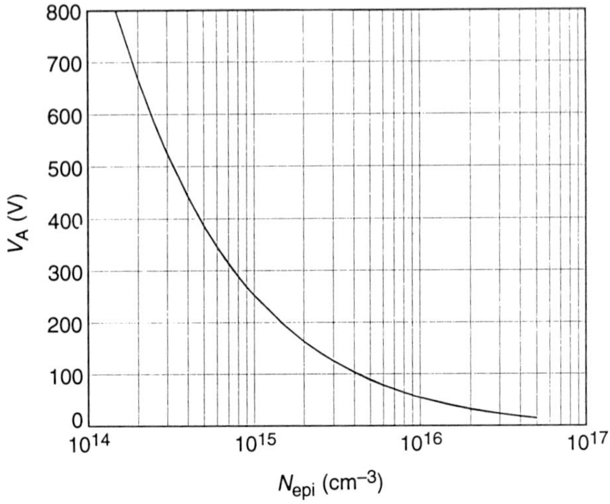

Figure 1.10. *Early voltage of the vertical npn transistor vs.* N_{epi}.

is plotted in Fig. 1.10 as a function of the epitaxial layer doping concentration. For proper transistor operation the value of V_{A} should be as large as possible. At an increasing epilayer doping concentration (= collector doping concentration) the collector-base depletion layer extends more into the base and modulations caused by changes in the collector-base reverse voltage cause basewidth modulation and thus affect the base transport efficiency to a larger extent, resulting in a reducing V_{A}.

The expression for the Early voltage in the lateral pnp can be derived in a similar way and yields:

$$V_{\mathrm{A,pnp}} = \sqrt{\frac{2qN_{\mathrm{epi}}{W_{\mathrm{B}}}^2(N_{\mathrm{epi}}+N_{\mathrm{B}})}{\epsilon_{\mathrm{Si}}N_{\mathrm{B}}}\varphi_{\mathrm{f,cb}}}, \tag{1.6}$$

Figure 1.11 shows a plot of the Early voltage of the lateral pnp transistor as a function of the epitaxial layer concentration and the collector-to-emitter mask separation, W_{M}. As the collector doping profile remains unchanged, the collector-base depletion layer extends mainly in the baselayer (=epilayer) in case of an

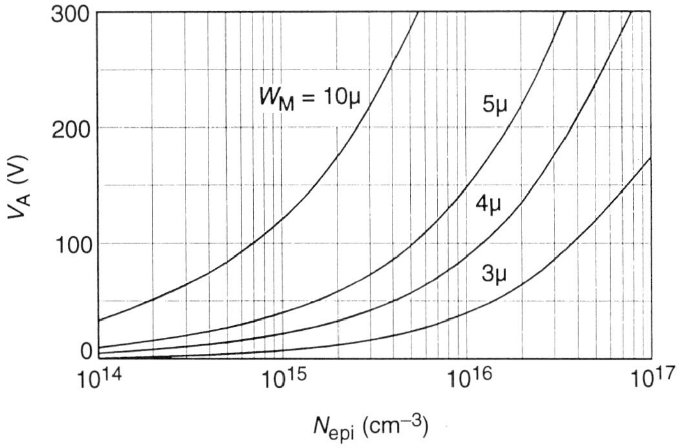

Figure 1.11. *Early voltage of the lateral pnp transistor vs.* N_{epi}.

epilayer with a low value of the doping concentration. Therefore, V_{A} increases with epitaxial layer doping concentration.

The switching time in saturated logic is limited by the time it takes to store or remove charge in the epilayer [1.12]. As indicated in Fig. 1.12, this storage time constant, τ_{s}, is a direct function of the epitaxial layer doping concentration. Common-emitter current gain and Early voltage primarily influence analog performance, whereas the storage time constant is of importance in switching circuits and non-saturating logic is to be used in case of large variations in epilayer thickness.

The epilayer width and doping concentration in a typical standard bipolar process are about 4μm and $5 \cdot 10^{15}$ cm^{-3} respectively. From the figures shown it can be seen that the nominal specifications are: $\beta_{\text{F,npn}} = 65$, $\beta_{\text{F,pnp}} = 55$, $V_{\text{br,npn}} = 15$ V, $V_{\text{br,pnp}} = 40$ V, $V_{\text{A,npn}} = 100$ V and $V_{\text{A,pnp}} = 15$ V (as minimum W_{M}). A departure from the standard processing clearly results in changes in these key specifications. Whether the process is compliant to the sensor processing-induced modifications can be evaluated directly from the figures.

Additional high-temperature processing results in further junction drive-in and lateral outdiffusion of dopants. This effect is es-

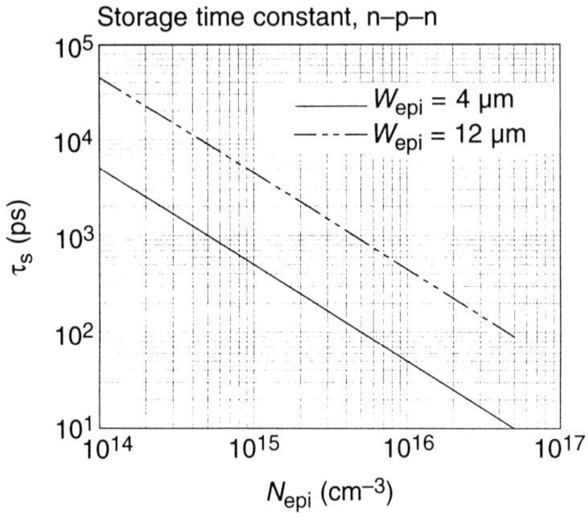

Figure 1.12. *Storage time constant vs.* N_{epi}.

pecially dominant in the lateral pnp, where the lateral diffusion reduces the basewidth and thus enhances β_F up to a limit where the transistor is not practically applicable and mask dimensions have to be increased.

The common-emitter current gain and the Early voltage of both the vertical npn and the lateral pnp transistor are shown in Fig. 1.13 as a function of epitaxial layer doping concentration; the effects of different post-processing temperature steps are shown.

This brief overview of a typical standard bipolar process indicates that difficulties are to be expected when changing technology parameters. Although some problems, such as breakdown in the lateral pnp at decreasing epilayer doping concentration, can be circumvented by adjusting design rules, there are clear limits to the range in which the epilayer width and doping concentration can be varied. Generally, the post-processing temperature load should be minimized.

1.4.2 CMOS processing

A typical CMOS process involves the fabrication of n-channel and p-channel enhancement transistors. Connecting these transistors

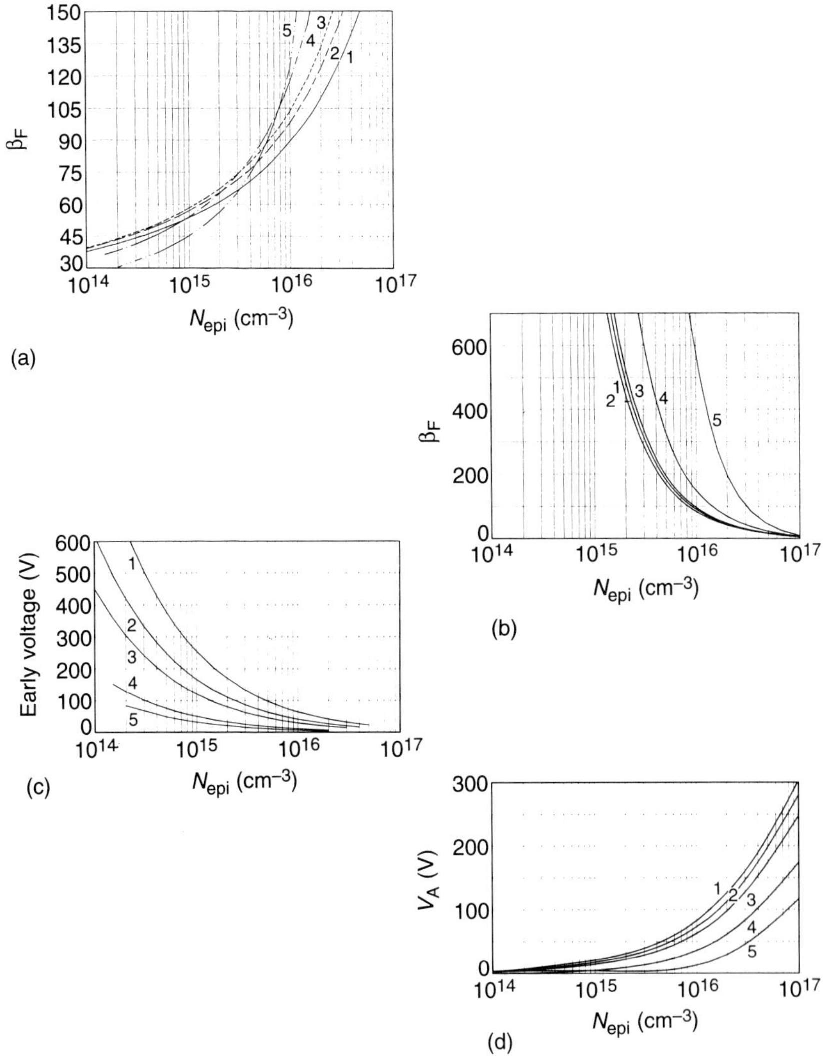

Figure 1.13. *(a) β_F of the npn transistor vs. N_{epi}, (b) β_F of the lateral pnp transistor vs. N_{epi}, (c) V_A of the npn transistor vs. N_{epi} and (d) V_A of the lateral pnp transistor vs. N_{epi}. Process (1) denotes standard processing, process (2) standard processing plus an additional anneal at 1000°C for 30 min, (3) 1000°C for 60 min, (4) 1100°C for 30 min and (5) 1100°C for 60 min.*

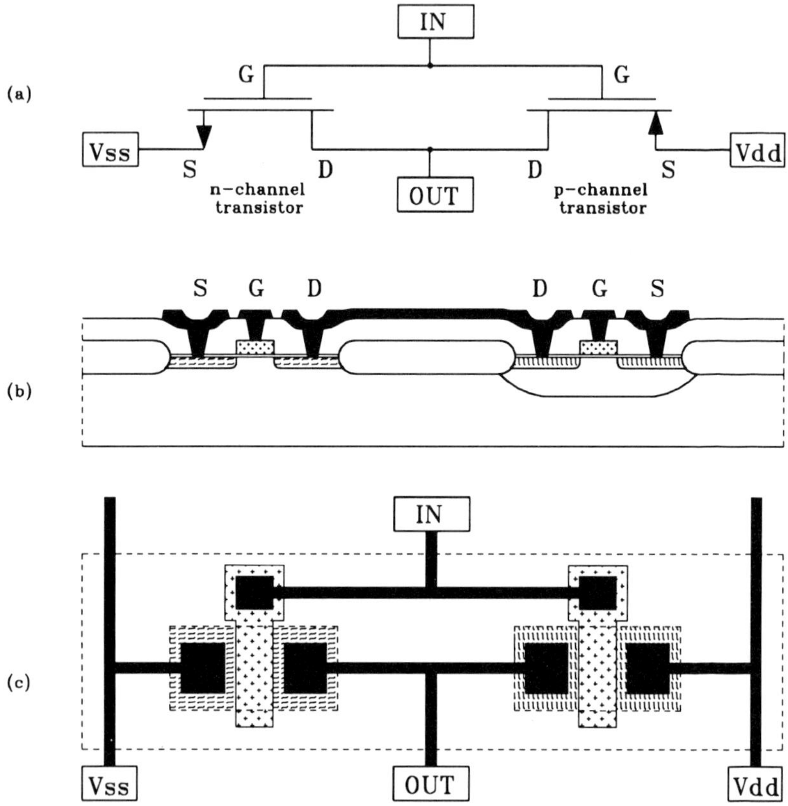

Figure 1.14. *Simple integrated CMOS invertor.*

in series results in the basic CMOS inverter shown in Fig. 1.14. As both the n-channel and p-channel transistor should be fabricated in the same substrate, an n-well is realized in a p-substrate. Ion implantation is typically used to ensure an n-channel device with a sufficient threshold voltage. The n-channel enhancement transistor is by definition non-conducting at $V_{gs} = 0$ ($V_i = V_{ss}$) and starts conducting at input voltages exceeding the threshold voltage, V_{Tn}. Alternatively, the p-channel transistor is non-conducting at $V_{gs} = 0$ V ($V_i = V_{dd}$) and starts conducting for input voltages smaller than $V_{dd} + V_{Tp}$. Therefore, for proper operation of the inverter, the threshold voltage of the n-channel transistor should be positive and that of the p-channel transistor should be negative. Typically $V_{Tn} \simeq 1$ V and $V_{Tp} \simeq -1$ V.

Fabrication starts with a p-type substrate on which a thin oxide layer and a nitride layer are subsequently deposited and patterned to define the borders for device isolation [1.8]. As the nitride film prevents access of oxygen to form silicon dioxide at the active areas, only the exposed areas are oxidized, as shown in Fig. 1.15a. This LOCOS (LOCal Oxidation of Silicon) oxide is usually preceeded by a boron channel stopping implant. The thin oxide film is required to ensure sufficient adhesion of the nitride film to the substrate. The nitride is removed after LOCOS and the parts where the n-well is to be formed are lithographically determined to result in the structure shown in Fig. 1.15b. Subsequently, the resist and the oxide film are stripped and the gate oxide is thermally grown. Usually a thin sacrificial oxide is first grown and etched in HF to leave a surface of optimum quality. After growth of the gate oxide of typically 500–1000 Å thickness, a phosphorus-doped polysilicon layer is deposited and patterned to form the gate. Doping takes place *in-situ* or using ion implantation after deposition. At this stage the device structure looks as shown in Fig. 1.15c.

The threshold voltage of the n-channel transistors has to be adjusted using an unmasked boron implantation. The implantation takes place through the gate and the gate oxide. Masking is not needed as the relatively small dose is highly overcompensated in other parts of the active area by either the source/drain implant or the n-well implant. As a next step the drain and source are to be realized. From Fig. 1.15c it can be seen that no masking is required within the transistor. The energy of the ion implantation is not sufficient to penetrate through the LOCOS and all the area between the gate and the LOCOS is intended for drain and source, which greatly facilitates alignment. After fabrication of the self-aligned n^+ drain and source in the n-channel transistor, while masking the p-channel transistors, and similarly also fabricating the p^+ drain and source in the p-channel transistors, a structure emerges as shown in Fig. 1.15d. A dielectric layer is subsequently deposited to enable isolation between the gate polysilicon (first polysilicon) and a second polysilicon or metal. Integrated capacitors are usually

(a)

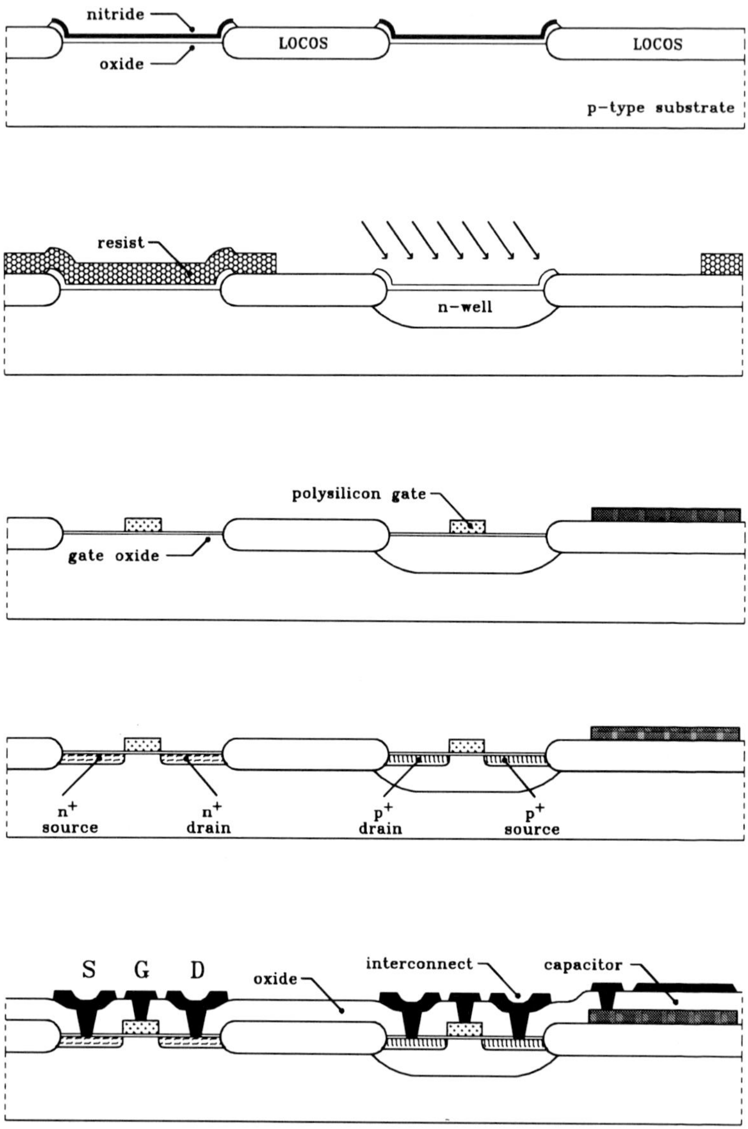

(b)

(c)

(d)

(e)

Figure 1.15. *CMOS process flow.*

poly1–poly2 capacitors. Finally, contact holes are etched and the second polysilicon or aluminum interconnect is deposited and patterned according to the electrical functioning of the integrated circuit and results in a structure as shown in Fig. 1.15e.

This highly simplified processing sequence clearly indicates the factors that limit the compliance of standard CMOS processing to integrated sensor fabrication-induced departures. The relevant factors are identified as: (1) the substrate doping, (2) the gate oxide, (3) the threshold voltage adjusting ion implantation and (4) the lateral diffusion of drain and source. These factors are affected by the following two sensor processing steps: (a) changing the substrate doping concentration and (b) thermal post-processing.

The substrate doping concentration directly determines the threshold voltage of the n-channel transistor. The threshold voltage of a MOS transistor is defined as:

$$V_{\mathrm{T}} = \varphi_{\mathrm{gs}} - 2\varphi_{\mathrm{f}} - \varphi_{\mathrm{ss}} - \varphi_{\mathrm{depl}} = \varphi_{\mathrm{gs}} - 2\varphi_{\mathrm{f}} - \frac{Q_{\mathrm{ss}}}{C_{\mathrm{ox}}} - \frac{Q_{\mathrm{depl}}}{C_{\mathrm{ox}}}, \quad (1.7)$$

with φ_{gs} the difference in work function between the gate material and the silicon, φ_{f} the electrostatic potential in silicon, φ_{ss} the potential charge due to surface charge, φ_{depl} the potential charge due to the charge stored in the depletion layer, Q_{ss} the positive surface charge in the gate oxide per unit area, Q_{depl} the charge stored in the depletion layer per unit area and C_{ox} the gate oxide capacitance per unit area. Assuming a polysilicon gate doping concentration equal to $N_{\mathrm{poly}} = 10^{19}\ \mathrm{cm}^{-3}$, a substrate doping concentration $N_{\mathrm{subs}} = 10^{15}\ \mathrm{cm}^{-3}$, a surface charge equal to $Q_{\mathrm{ss}} = q \times N_{\mathrm{ss}} = 16 \cdot 10^{-20} \times 5 \cdot 10^{10} = 8 \cdot 10^{-9}\ \mathrm{Ccm}^{-2}$ [1.13] and $C_{\mathrm{ox}} = \epsilon_{\mathrm{ox}}/t_{\mathrm{ox}} = 3.9 \times 8.85 \cdot 10^{-14}/10^{-5} = 35 \cdot 10^{-9}\ \mathrm{Fcm}^{-2}$ (gate oxide $t_{\mathrm{ox}} = 1000$ Å) with ϵ_{ox} the oxide dielectric permittivity results in:

$$\begin{aligned} \varphi_{\mathrm{gs}} &= \varphi_{\mathrm{f}}(p - \mathrm{subs}) - \varphi_{\mathrm{f}}(n - \mathrm{polygate}) \\ &= \frac{kT}{q} \ln\left(\frac{{n_{\mathrm{i}}}^2}{N_{\mathrm{poly}} N_{\mathrm{subs}}}\right) \\ &= 0.026 \ln\left(\frac{\left(1.45 \cdot 10^{10}\right)^2}{10^{19} \cdot 10^{15}}\right) = -0.82\ \mathrm{V} \end{aligned}$$

$$\begin{aligned}
\varphi_{\rm f}(p-\text{subs}) &= \frac{kT}{q}\ln\left(\frac{n_{\rm i}}{N_{\rm subs}}\right) = 0.026\ln(1.45\cdot 10^{-5}) \\
&= -0.29\ \text{V} \qquad (1.8) \\
\varphi_{\rm ss} &= \frac{Q_{\rm ss}}{C_{\rm ox}} = 0.23\ \text{V} \\
\varphi_{\rm depl} &= \frac{Q_{\rm depl}}{C_{\rm ox}} = -\frac{\sqrt{2q\epsilon_{\rm Si}N_{\rm subs}(-)2\varphi_{\rm f}}}{C_{\rm ox}} \\
&= -\frac{\sqrt{32\cdot 10^{-20}\times 11.8\times 8.85\cdot 10^{-14}\times 10^{15}\times 0.58}}{35\cdot 10^{-9}} \\
&= -0.40\ \text{V}.
\end{aligned}$$

Thus $V_{\rm Tn} = -0.82+0.58-0.23+0.40 = -0.07$ V. Ion implantation should be used to adjust the threshold voltage. To achieve $V_{\rm Tn} =$ 1 V a boron implantation is required with a dose equal to: $\Delta V_{\rm T}\times C_{\rm ox}/q = 1.07\times 35\cdot 10^{-9}/16\cdot 10^{-20} = 23\cdot 10^{10}$ cm^{-2} [1.8]. However, the required dose changes with substrate doping concentration. If it is assumed that sensor processing imposes $N_{\rm subs} = 2\cdot 10^{16}$ cm^{-3}, this non-standard substrate results in a threshold voltage equal to:

$$\begin{aligned}
\varphi_{\rm gs} &= \varphi_{\rm f}(p-\text{subs}) - \varphi_{\rm f}(n-\text{polygate}) = \\
&= 0.026\ln\left(\frac{(1.45\cdot 10^{10})^2}{10^{19}\cdot 2\cdot 10^{16}}\right) = -0.90\ \text{V} \\
\varphi_{\rm f}(p-\text{subs}) &= \frac{kT}{q}\ln\left(\frac{n_{\rm i}}{N_{\rm subs}}\right) \\
&= 0.026\ln\left(\frac{1.45\cdot 10^{10}}{2\cdot 10^{16}}\right) = -0.37\ \text{V} \\
\varphi_{\rm ss} &= \frac{Q_{\rm ss}}{C_{\rm ox}} = 0.23\ \text{V} \qquad (1.9) \\
\varphi_{\rm depl} &= \frac{Q_{\rm depl}}{C_{\rm ox}} = \frac{-\sqrt{2q\epsilon_{\rm Si}N_{\rm subs}(-)2\varphi_{\rm f}}}{C_{\rm ox}} = -2.01\ \text{V} \\
V_{\rm Tn} &= -0.90+0.74-0.23+2.01 = 1.61\ \text{V}
\end{aligned}$$

The threshold voltage at a substrate doping concentration exceeding 10^{16} cm^{-3} clearly cannot be adjusted to the target value using boron implantation. However, the effect for substrate doping concentration within the range between 10^{15} and 10^{16} cm^{-3} can in

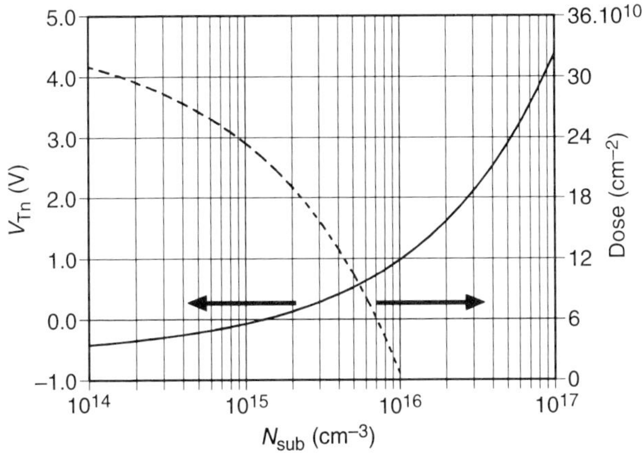

Figure 1.16. V_{Tn} *vs. substrate doping concentration.*

principle be compensated. Figure 1.16 shows the n-channel threshold voltage as a function of substrate doping concentration and also the boron implantation dose as a function of N_{subs} that would be required to maintain $V_{\mathrm{Tn}} = 1$ V.

However, for proper operation of the CMOS circuits the p-channel transistors also should have the correct threshold voltage; $V_{\mathrm{Tp}} = -1$ V. This threshold voltage is directly determined by the n-well doping concentration. The n-well doping concentration should increase with substrate doping concentration to ensure reproducibility and to have the well-substrate depletion layer extend mainly into the substrate in order to limit the body effect. Assuming a nominal doping concentration in the n-well equal to $N_{\mathrm{well}} = 2 \cdot 10^{15}$ cm^{-3} results in:

$$\begin{aligned}
\varphi_{\mathrm{gs}} &= \varphi_{\mathrm{f}}(n - \mathrm{well}) - \varphi_{\mathrm{f}}(n - \mathrm{polygate}) \\
&= \frac{kT}{q} \ln\left(\frac{N_{\mathrm{well}}}{n_{\mathrm{poly}}}\right) = \\
&= 0.026 \ln\left(\frac{2 \cdot 10^{15}}{10^{19}}\right) = -0.22\ \mathrm{V}
\end{aligned}$$

$$\begin{aligned}
\varphi_{\mathrm{f}}(n-\mathrm{well}) &= \frac{kT}{q}\ln\left(\frac{N_{\mathrm{well}}}{n_{\mathrm{i}}}\right) = 0.026\ln\left(\frac{2\cdot 10^{15}}{1.45\cdot 10^{10}}\right) \\
&= 0.31\ \mathrm{V} \\
\varphi_{\mathrm{ss}} &= \frac{Q_{\mathrm{ss}}}{C_{\mathrm{ox}}} = 0.23\ \mathrm{V} \qquad (1.10) \\
\varphi_{\mathrm{depl}} &= \frac{Q_{\mathrm{depl}}}{C_{\mathrm{ox}}} = \frac{\sqrt{2q\epsilon_{\mathrm{Si}}N_{\mathrm{well}}2\varphi_{\mathrm{f}}}}{C_{\mathrm{ox}}} \\
&= \frac{\sqrt{32\cdot 10^{-20}\times 11.8\times 8.85\cdot 10^{-14}\times 2\cdot 10^{15}\times 0.62}}{35\cdot 10^{-9}} \\
&= 0.58\ \mathrm{V} \\
V_{\mathrm{Tp}} &= -0.22-0.62-0.23-0.58 = -1.65\ \mathrm{V}
\end{aligned}$$

This value is within the range that is practically applicable in a CMOS process. When increasing the substrate doping concentration to $N_{\mathrm{subs}} = 10^{16}\ \mathrm{cm}^{-3}$, the well doping should be at least $N_{\mathrm{well}} = 2\cdot 10^{16}\ \mathrm{cm}^{-3}$, resulting in a threshold voltage equal to:

$$\begin{aligned}
\varphi_{\mathrm{gs}} &= \varphi_{\mathrm{f}}(n-\mathrm{well}) - \varphi_{\mathrm{f}}(n-\mathrm{polygate}) \\
&= 0.026\ln\left(\frac{2\cdot 10^{16}}{10^{19}}\right) = -0.16\ \mathrm{V} \\
\varphi_{\mathrm{f}}(n-\mathrm{well}) &= \frac{kT}{q}\ln\left(\frac{N_{\mathrm{well}}}{n_{\mathrm{i}}}\right) \\
&= 0.026\ln\left(\frac{2\cdot 10^{16}}{1.45\cdot 10^{10}}\right) = 0.37\ \mathrm{V} \\
\varphi_{\mathrm{ss}} &= \frac{Q_{\mathrm{ss}}}{C_{\mathrm{ox}}} = 0.23\ \mathrm{V} \qquad (1.11) \\
\varphi_{\mathrm{depl}} &= \frac{Q_{\mathrm{depl}}}{C_{\mathrm{ox}}} = \frac{\sqrt{2q\epsilon_{\mathrm{Si}}N_{\mathrm{well}}2\varphi_{\mathrm{f}}(\mathrm{well})}}{C_{\mathrm{ox}}} = 2.00\ \mathrm{V} \\
V_{\mathrm{Tp}} &= -0.16-0.74-0.23-2.00 = -3.13\ \mathrm{V}
\end{aligned}$$

The threshold voltage of the p-channel transistor as a function of n-well doping concentration is shown in Fig. 1.17. This value of the threshold voltage cannot be used in a 5 V CMOS process. The influence of substrate doping concentration of threshold voltage can be circumvented by using the twin well structure shown in Fig. 1.18 [1.8]. A lightly doped epilayer is grown on a substrate of any

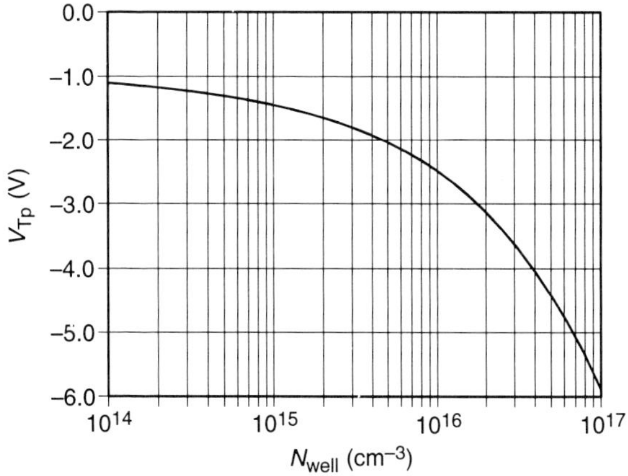

Figure 1.17. V_{Tp} *vs.* N_{well}.

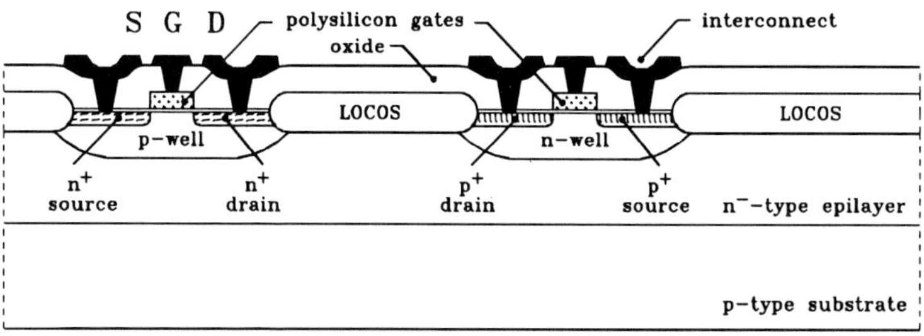

Figure 1.18. *Twin-well structure.*

doping type and concentration. A p-well is defined to fabricate the n-channel transistor and similarly an n-well is implanted to form the active area of the p-channel transistor. Obviously, the well concentration can be adjusted to match the threshold requirements of the transistors, irrespective of the substrate doping concentration, but at the expense of an increased processing complexity.

The second fabrication compatibility infringement is due to the gate oxide. In particular the diffusion of phosphorus from the

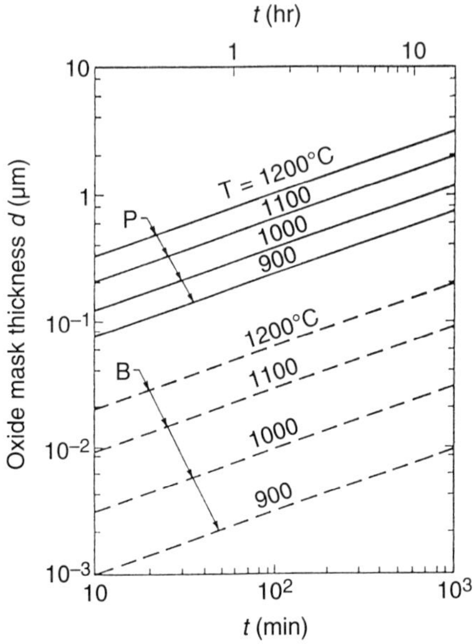

Figure 1.19. *Minimum masking oxide thickness.*

polysilicon gate through the gate oxide into the channel could degrade device performance. From a technology point of view the highly doped polysilicon gate can be considered as a deposited dopant and the post-processing temperature step as the drive-in. The gate oxide acts in principle as a masking material that would prevent drive-in at the channel area. In practice, however, conventional dopants such as boron and phosphorus have a finite diffusion in silicon dioxide and thus a minimum oxide thickness is required, as shown in Fig. 1.19, to ensure sufficient masking [1.8]. Clearly, the gate oxide thickness $t_{ox} = 1000$ Å is insufficient to prevent phosphorus incorporation into the channel. The phosphorus dose would result in a negative shift in the threshold voltage.

Diffusion of phosphorus through silicon dioxide depends on temperature and oxide quality, where typically, $5 \cdot 10^{-19} < D_{p,ox} < 5 \cdot 10^{-15}$ cm^{2}s^{-1}. Assuming a diffusion coefficient of phosphorus in silicon dioxide $D_{p,ox} \simeq 10^{-15}$ cm^{2}s^{-1} and $D_{p,Si} \simeq 10^{-15}$ cm^{2}s^{-1} (1000°C) and a number of impurities in a 0.5 μm thick polysilicon layer equal to $Q = N_{poly} \times t_{poly} = 10^{19} \times 5 \cdot 10^{-5} = 5 \cdot 10^{14}$ cm^{-2}

gives for the concentration at the oxide–silicon interface, N, and the junction, x_j:

$$
\begin{aligned}
N(1000\,\text{Å}, 1\,\text{hr}) &= \frac{Q}{\sqrt{\pi D_{\mathrm{p,ox}} t}} \exp\left(\frac{-{t_{\mathrm{ox}}}^2}{4 D_{\mathrm{p,ox}} t}\right) \\
&= \frac{5 \cdot 10^{14}}{\sqrt{\pi \cdot 10^{-15} \cdot 3600}} \exp\left(\frac{-10^{-10}}{4 \cdot 10^{-15} \cdot 3600}\right) \\
&= 33 \cdot 10^{17}\,\mathrm{cm}^{-3} \\
x_j + t_{\mathrm{ox}} &= \sqrt{4 D_{\mathrm{p,Si}} \ln\left(\frac{Q}{N_{\mathrm{subs}} \sqrt{\pi D_{\mathrm{p,Si}} t}}\right)} \qquad (1.12) \\
&= \sqrt{144 \cdot 10^{-13} \ln\left(\frac{5 \cdot 10^{14}}{10^{15} \sqrt{36\pi 10^{-13}}}\right)} \\
&= 0.13\,\mu m
\end{aligned}
$$

Thus, the dose is about equal to $14 \cdot 10^{16} \times 13 \cdot 10^{-6} = 182 \cdot 10^{10}$ cm^{-2}, which gives rise to a threshold shift about equal to:

$$\Delta V_{\mathrm{T}} = -q \times \mathrm{dose}/C_{\mathrm{ox}} = -16 \cdot 10^{-20} \times 182 \cdot 10^{10}/35 \cdot 10^{-9} = -8.5\ \mathrm{V}.$$

Therefore, the phosphorus dose in the channel due to high-temperature post-processing steps on poor-quality gate oxide causes an unacceptable negative shift of the threshold voltage in both the n-channel and the p-channel transistor. For $D_{\mathrm{p,ox}} < 10^{-16}$ $\mathrm{cm^2 s^{-1}}$ the effect can be disregarded. This effect is, at a specified thickness of the gate oxide, completely dependent on CMOS process control.

Another disadvantageous effect resulting from high-temperature processing subsequent to standard CMOS microelectronic fabrication is the lateral diffusion of the drain and source dopants. CMOS processing ensures self-alignment of drain and source with respect to the gate with minimum lateral diffusion, so that the gate capacitance is mainly composed of the gate-body capacitance, C_{gb}, as shown in Fig. 1.20. However, an increasing subsequent thermal loading results in lateral diffusion and thus in an increasing drain-gate and source-gate overlap and consequently in an increasing drain-gate and source-gate capacitance.

In the case of the n-channel transistor with $N_{\mathrm{subs}} = 10^{15}$ cm^{-3}

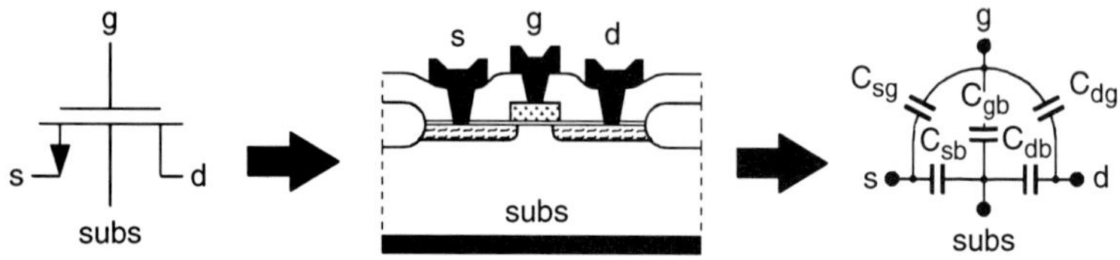

Figure 1.20. *Parasitic capacitances in a CMOS transistor.*

and $N_{\text{drain}} = N_{\text{source}} \simeq 10^{20}$ cm^{-3}, a thermal treatment at 1000°C for 60 min would result in channel narrowing due to lateral diffusion of the phosphorus doping profile equal to:

$$\begin{aligned} \Delta x &= 2\sqrt{Dt}\,\text{erfc}^{-1}\left(\frac{N_{\text{subs}}}{N_{\text{drain}}}\right) \\ &= 5.4\sqrt{Dt} = 5.4\sqrt{10^{-15} \times 3600} = 0.1\ \mu\text{m}. \end{aligned} \tag{1.13}$$

The change in drain-gate capacitance per unit of width of the gate is $\Delta C_{\text{dg}} = \epsilon \Delta x / t_{\text{ox}} = 345$ fFcm^{-1}. The drain-gate capacitance has a much larger impact on the high-frequency performance of an integrated amplifier than the gate-body capacitance, due to the Miller effect, which causes the amplifier to exhibit an apparent capacitance proportional to the voltage gain and C_{dg}. The effect is somewhat reduced when using a lightly doped drain that is conventionally used to reduce the electric field in the drain region in order to prevent hot electrons from charging the gate oxide.

So far the limits of the allowable post-processing based on the compliance of standard bipolar and CMOS processing have been discussed. At this stage it is important to investigate whether such changes are necessary in practical silicon sensor types that are generally expected to have a large potential for realization as an integrated silicon smart sensor.

1.5 Transduction effects in silicon for on-chip integrated sensors

The bipolar and CMOS processes discussed in the previous section are more or less compliant to sensor process-induced modifications in the standard processing. The compliance is the major issue in functional integration rather than integration density. The sensor dimensions usually exceeds the die area required for the integrated circuit anyway. The requirements imposed by the various non-electrical signal domains on the fabrication compatibility and the most promising transduction effects in view of their suitability to be merged with bipolar or CMOS processing will be discussed in this section.

1.5.1 Thermal sensors

The thermal domain is the least demanding non-electrical signal domain with respect to fabrication compatibility. Silicon exhibits profound temperature effects, such as the temperature dependence of the junction forward voltage drop, due to the temperature dependence of the bandgap, and the Seebeck effect. The temperature dependence of the junction forward voltage is usually regarded as a nuisance in analog circuit design and special measures are taken to maintain proper operation of the circuit over a wide temperature range. However, this temperature dependence; $V_{\mathrm{f}} = V_{\mathrm{T}} \ln(I_{\mathrm{f}}/I_{\mathrm{s}})$, where $V_{\mathrm{T}} = kT/q$ and I_{s} denotes the saturation current, can also be used to realize integrated thermal sensors without any special sensor processing. The highly simplified schematic diagram of such a Proportional To Absolute Temperature circuit (PTAT) is depicted in Fig. 1.21a [1.14]. The sensor supplies an output current:

$$I(T) = \frac{k}{qR} \ln{(m)} T \tag{1.14}$$

and is suitable for integration in any bipolar process that provides transistors of acceptable performance for analog integrated circuits. The only fabrication compatibility problem arising in this type of integrated silicon smart temperature sensor is self-heating of the chip due to internal power dissipation in combination with

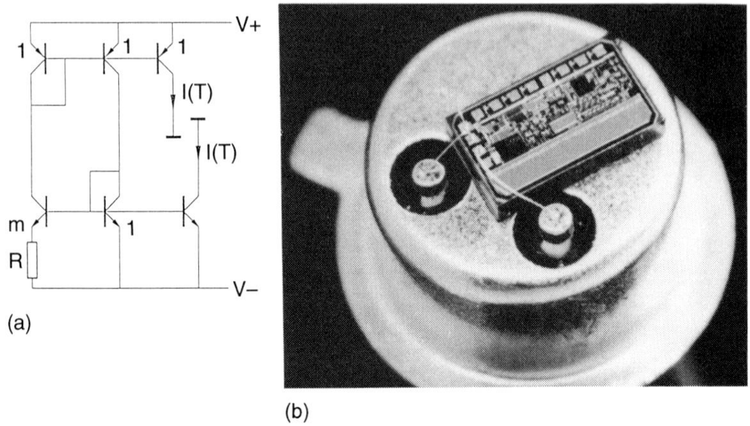

Figure 1.21. *(a) Simplified schematic diagram of the PTAT circuit. (b) Photograph of a commercial integrated silicon smart thermal sensor.*

the limited heat conductance of the conventional molded plastic package. Self-heating would lead to a measurement error equal to $\Delta T = P_{\mathrm{diss}} R_{\mathrm{th}}$ and directly determines the sensor inaccuracy specification. This effect can be minimized using non-standard packaging or low-power circuits. A photograph of an integrated silicon smart temperature is shown in Fig. 1.21b.

The second well-established temperature effect in silicon is the Seebeck effect [1.15]. Integrated thermopiles can be fabricated by placing a number of thermocouples electrically in series and thermally in parallel over a temperature gradient on the chip as shown in Fig. 1.22a. A voltage is generated proportional to the temperature difference and the number of thermocouples. Thermocouples with reasonably high Seebeck coefficient (up to 1 $\mathrm{mVK}^{-1} \times$ junction) can be realized using silicon–aluminum junctions or junctions between silicon layers of different doping. The fabrication of integrated thermopiles is, therefore, directly compatible with standard bipolar processing. However, integrated silicon thermopiles cannot be practically applied without additional special sensor processing. The high thermal conductance of silicon prevents a temperature gradient being built up by internal or

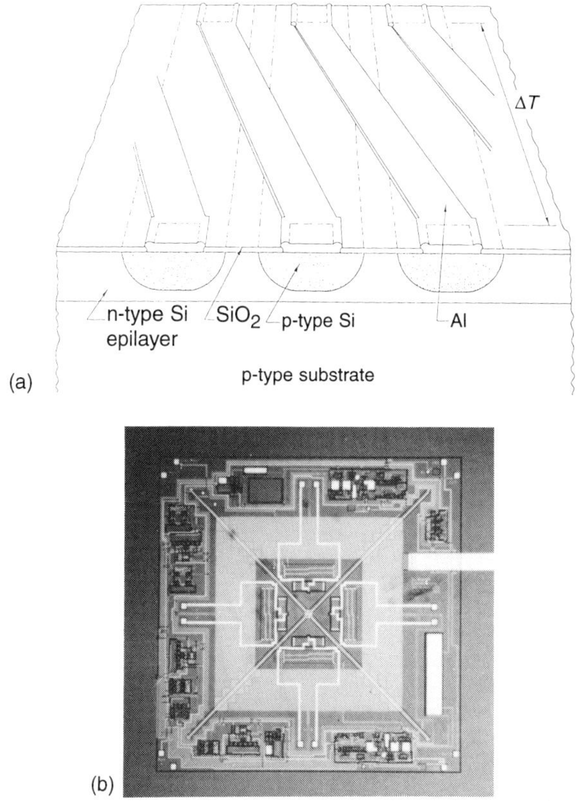

Figure 1.22. *(a) Integrated thermopile in silicon. (b) Integrated silicon smart thermal flow sensor using the flow-induced temperature gradient over a heated chip surface.*

external heating sources of reasonable power. The wafer has to be subjected to bulk micromachining after completion of the standard bipolar processing before the integrated smart sensor is suitable for applications such as thermal infrared detection or thermal flow sensing (Fig. 1.22b). The extent of the fabrication compatibility problem for thermopile-based sensors is, therefore, determined by the compatibility of micromachining and packaging technologies with microelectronic processing, which will be discussed in the next chapter.

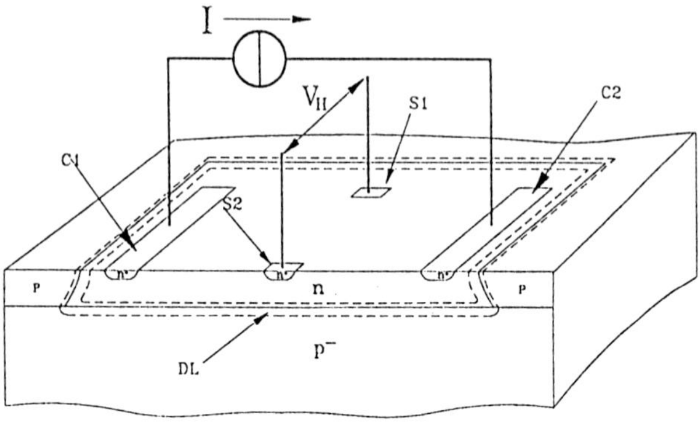

Figure 1.23. *Integrated silicon Hall plate.*

1.5.2 Magnetic sensors

Sensors for magnetic signals are also relatively easy to merge with a bipolar process. The most popular types of silicon magnetic sensors are the Hall plate and the dual-collector magnetotransistor [1.16]. Figure 1.23 shows a typical integrated Hall plate. Electric current flows in between the current contacts C_1 and C_2. In the case of a magnetic field normal to the silicon surface, the moving charge carriers experience a deflecting Lorentz force, which gives rise to a Hall field perpendicular to the direction of the current. The resulting Hall voltage is measured over the sensor contacts S_1 and S_2 as:

$$V_{\mathrm{H}} = \frac{r_{\mathrm{H}} G}{q N_{\mathrm{H}} t_{\mathrm{H}}} IB, \tag{1.15}$$

where r_{H} denotes the Hall constant, G a geometric factor, B the magnetic field, N_{H} the layer doping concentration and t_{H} the thickness of the Hall plate. Obviously, the sensitivity is maximized when the factor $N_{\mathrm{H}} \cdot t_{\mathrm{H}}$ is minimized. In a bipolar process $N_{\mathrm{H}} = N_{\mathrm{epi}}$ and $t_{\mathrm{H}} = t_{\mathrm{epi}} - DL$. The typical standard bipolar process used here with $t_{\mathrm{H}} \simeq t_{\mathrm{epi}}$ results in a sensitivity $S = r_{\mathrm{H}}/(q \cdot N_{\mathrm{epi}} \cdot t_{\mathrm{epi}}) = 150$ $\mathrm{VA^{-1}T^{-1}}$, which increases linearly with decreasing epitaxial layer doping concentration. However, when reducing N_{epi} below a certain level, the assumption $t_{\mathrm{H}} = t_{\mathrm{epi}}$ does not hold due to the increasing

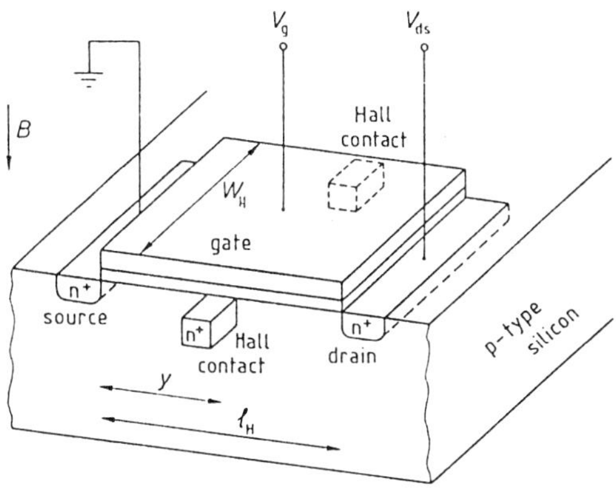

Figure 1.24. *CMOS Hall plate.*

epi-substrate depletion layer thickness extending into the epilayer. The thickness of the depleted epilayer equals:

$$DL = \sqrt{\frac{2\epsilon V_{es}}{q} \frac{N_{epi}}{N_{subs} \cdot (N_{epi} + N_{subs})}}. \tag{1.16}$$

Moreover, t_H becomes susceptible to changes in V_{es}, which directly limits the inaccuracy specification.

Generally, the indicated sensitivity is sufficient, as the limit of detectivity is set by offset. The mere sensor requirements give little reason to depart from the standard bipolar process, which, conveniently, also ensures fabrication compatibility on the wafer level. Hall plates have also been merged with CMOS processes. An example is shown in Fig. 1.24 [1.17]. The Hall contacts are directly fabricated in the substrate, which is of a suitable doping concentration in the standard process ($N_{subs} = 10^{15}$ cm^{-3}) to give an adequate sensitivity.

The fabrication compatibility problem arises in the Hall sensor on the die level during packaging. Encapsulation stress is introduced during cool-down after die bonding of the silicon chip on a material of different thermal expansion coefficient or after packaging. The piezoresistive effect causes a sheet resistivity non-uniformity in the Hall plate, which in turn results in a Hall voltage offset. This

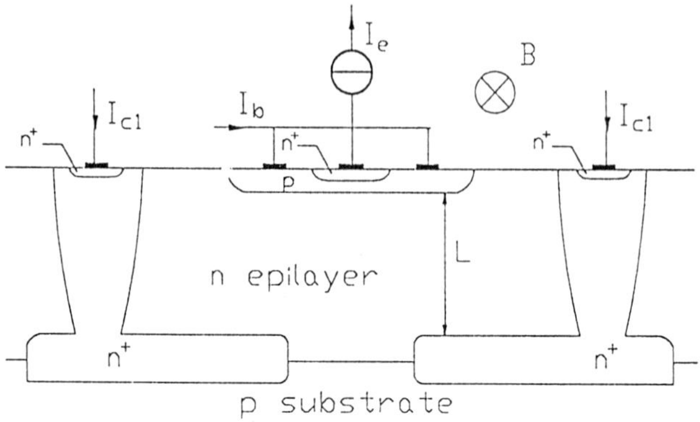

Figure 1.25. *Dual-collector magnetotransistor.*

offset can in principle be compensated during calibration; however, offset drift due to creep in the adhesive material determines the detection limit. The effect of the encapsulation stress is minimized by aligning the Hall plate along the crystal axis with the minimum piezoresistive effect and by using bulk micromachining techniques to fabricate suspension structures to separate the Hall plate from the bonded area [1.18].

The dual-collector magnetotransistor shown in Fig. 1.25 is the second type of magnetic field sensor with sufficient potential to be considered for realization as an integrated silicon smart sensor in a bipolar process. Charge carriers are injected from the emitter into the base in which a fraction recombine. An increasing part of this base current flows, at an increasing magnetic field of the indicated polarity, from the lefthand extrinsic base, resulting in an increasing emitter-base localized forward voltage drop and, consequently, leads to increased injection at the lefthand of the emitter-base junction [1.19]. The resulting asymmetry in collector current, $\Delta I_c/I_e = S \cdot B$, with $S \simeq 10\%$ as a sensitivity factor, is measured. Recombination in the collector should be avoided, which implies depletion of the entire epilayer underneath the junction. About 2 μm of epitaxial layer thickness remains in between the buried layer and the extrinsic base layer, resulting in $V_{es} > (2\mu m)^2 \cdot qN_{epi}/(2\epsilon) = 32$ V. Lower values of this reverse voltage are required to avoid breakdown, therefore, a lower epitaxial doping concentration is preferred in this type of device.

Reducing the epilayer doping concentration to $N_{\mathrm{epi}} = 1 \cdot 10^{15}$ cm^{-3} results in the active device performance indicated in Figs 1.6–1.11; $\beta_{\mathrm{F,npn}} = 54$ instead of 65 and $V_{\mathrm{A,npn}} = 250$ instead of 15. Similarly, $\beta_{\mathrm{F,pnp}} = 65$ instead of 55 and $V_{\mathrm{A,pnp}} = 2.5$ instead of 15. The performance of the lateral pnp is, therefore, seriously impaired and the collector-emitter mask spacing has to be increased. The conventional molded plastic package does not significantly interfere with magnetic fields and does thus not add to the fabrication compatibility problem.

1.5.3 Radiant sensors

Although most radiant sensors are optical devices operating in the visible part of the spectrum, this category extends over the entire electro-magnetic spectrum and includes, in principle, microwave devices and X-ray detectors. Silicon demonstrates relatively strong photoelectric effects. The photojunction and photoresistance effects are the most notable. A characteristic of radiant sensors is the need for a window in front of the sensitive area, which is transparent for the part of the spectrum under consideration. Rather akin to the thermoelectric effect, impinging light is a nuisance in circuit design, as it adds to the leakage current in active devices. This observation suggests a fabrication incompatibility in the packaging phase of integrated silicon radiant smart sensors, unless special measures are taken during wafer processing to prevent active devices from being illuminated as well.

Research on high-density CCD imaging sensors could in a way be considered as the smart optical sensor field with the oldest tradition, as readout and charge transportation circuits are integrated. However, this statement is not quite correct, as special fabrication techniques are applied; no effort has been made to obtain fabrication compatibility with a standard process, but a special process for integration of readout circuits has been designed instead. Nevertheless, many of the solutions originating from this field, such as the division of the die into a sensitive area and a readout area covered by an opaque film, are highly suitable for

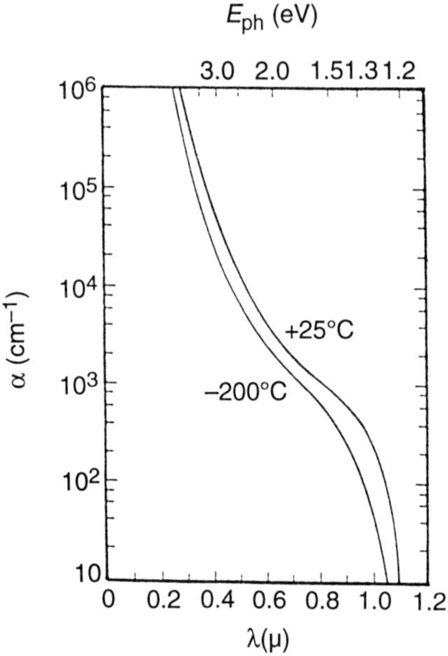

Figure 1.26. *Absorption coefficient in silicon vs. λ.*

adoption in integrated optical smart sensors to relieve the above-mentioned compatibility problem.

Incident photons of sufficient energy larger than the bandgap energy, $E_g = 1.12$eV, thus having a wavelength $\lambda < 1100$ nm are able to initiate bandgap crossings and to generate electron-hole pairs. When such an effect takes place at a pn junction, the electron-hole pairs are separated by the built-in electric field and the open-circuit potential is modulated or, when short-circuiting the junction, an external current becomes available. When the electron-hole generation takes place in a slab of doped silicon, excess charge carriers become available and the sheet resistivity of that layer is lowered. A common complication in silicon optical detector is the wavelength dependence of the absorption coefficient in the visible part of the spectrum, resulting in a penetration depth ranging between about 0.1 μm at λ=400 nm and about 10 μm at λ=650 nm as shown in Fig. 1.26 [1.20]. A typical photojunction detector relies on the diffusion of shallowly generated minority charge carriers, generated by short-wavelength light down

(a)

(b)

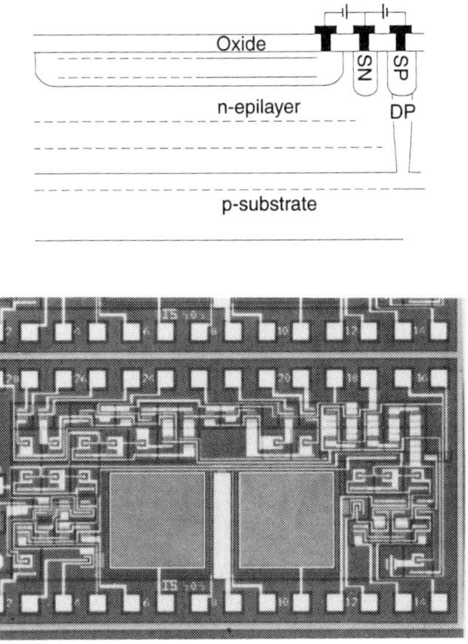

Figure 1.27. *(a) structure of the integrated silicon color sensor. (b) Photograph of a realized device.*

to the junction and bulk generated charge carriers, resulting from long-wavelength light, to diffuse upward. However, due to the limited lifetime of minority charge carriers and the junction depth, which is about 0.5–1 μm when using bipolar process-compatible layers, the spectral response of silicon photodiodes is wavelength dependent with a maximum at about $\lambda = 550$ nm, which is acceptable in general-purpose photodetectors. In a photoconductor this wavelength dependence of the absorption coefficient determines the minimum layer thickness.

In some optical devices, such as the integrated silicon color sensor depicted in Fig. 1.27a, special sensor requirements force departures from the standard bipolar process [1.20–1.21]. Sensor operation is based on the wavelength dependence of the absorption coefficient mentioned and a well-defined photocharge collection volume is established, by depleting the upper part of a silicon layer down from a shallow junction and the complementary part up from a lower junction. Applying a small reverse voltage across the upper junction, while maintaining a sufficiently high reverse voltage across the lower junction to achieve complete depletion of the epilayer in between, results in a superficial absorbing layer and thus in a predominantly short-wavelength (blue) response. Increasing the reverse voltage across the shallow junction, while adjusting that over the epilayer-substrate junction, until pinch-off in an integrated JFET takes place, results in an enhanced long-wavelength (red) response and enables the electronic tuning of the spectral response.

A 10 μm thick epitaxial layer is required to enable the efficient electronic control of the response up to wavelengths of about 600 nm. The epitaxial doping has to be lowered from that used in the standard bipolar process to about $1 \cdot 10^{15}$ cm^{-3} to enable full depletion of the epilayer with moderate values of the reverse voltage. Merging of the color sensor with the bipolar processes, therefore, imposes the following modifications: t_{epi} from 4 μm to 10 μm and N_{epi} from $5 \cdot 10^{15}$ cm^{-3} to $1 \cdot 10^{15}$ cm^{-3}. As can be derived from Figs 1.7–1.12, such changes have considerable impact on the static and dynamic performance of integrated active devices. Nevertheless, integrated silicon smart color sensors have been realized in the bipolar process indicated and the result is shown in Fig. 1.27b. Adjustment of a number of processing steps (epitaxial growth conditions and deep-boron isolation drive-in) and design rules (lateral pnp collector-emitter spacing) are required to obtain a sensor and active components of sufficient, although sub-optimum, performance.

The absorption coefficient in silicon for high-energy radiation is in the $1–10^{-2}$ cm^{-1} range. Therefore, even larger photon absorption/charge collection layers are required. In a practical microstrip radiation detector shown in Fig. 1.28, the full 4″ wafer thickness of 525 μm is depleted [1.22]. This principle is fundamentally incompatible with standard bipolar processing. However, it is exactly this device which would benefit greatly from integration as a smart sensor. The strip detector is a linear array sensor with over 1000 strips per wafer and on-chip multiplexing circuits strongly reduce the output connection problem. Moreover, each strip can be regarded as a current source with a relatively small capacitance and on-chip readout circuitry largely reduces the susceptibility to stray capacitances that would otherwise attenuate the signal. Therefore, this application benefits especially from the added value criterion to be introduced in chapter 4.

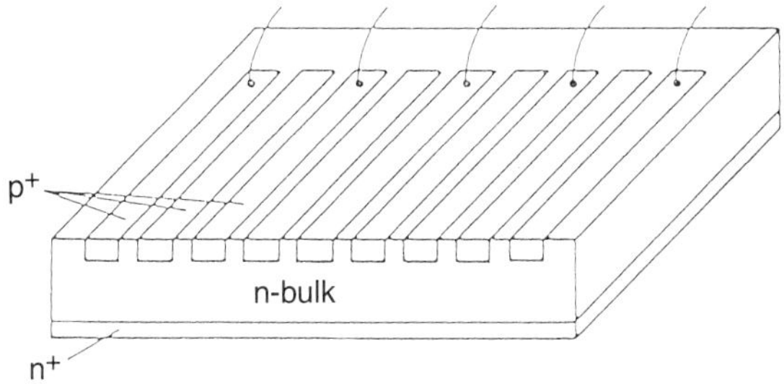

Figure 1.28. *Structure of the integrated silicon microstrip nuclear detector.*

1.5.4 Mechanical sensors

Mechanical sensors rely on the micromachining technologies that have been developed to sculpture micromechanical structures, such as membranes, cantilevers and beams. Two techniques are widely employed for this purpose; bulk micromachining and surface micromachining. These both have serious implications with respect to fabrication compatibility and will be discussed in chapter 2. Bulk micromachining involves the removal of selected parts of the bulk silicon using either isotropic etching, with an equal etch rate in each direction, or anisotropic etching, with a preferential etch with a stop at one of the crystal planes. Generally, anisotropic etching is preferred in micromachining for obtaining fine feature size. Automatic etch stop techniques are essential for obtaining thin membranes of high uniformity after etching from the backside through the entire wafer thickness using a dielectric window.

The deflection of the center of the bulk-micromachined circular membrane, shown in Fig. 1.29, can for an internal stress $\sigma < E(h/a)^2$ and a diameter, a, much larger than the deflection be expressed as a function of an applied pressure, P, as [1.21]:

$$W = 0.662 \cdot a \sqrt[3]{\frac{Pa}{Eh}} \tag{1.17}$$

This equation indicates that the sensitivity of deflection to pressure is mainly determined by membrane radius, which can be designed

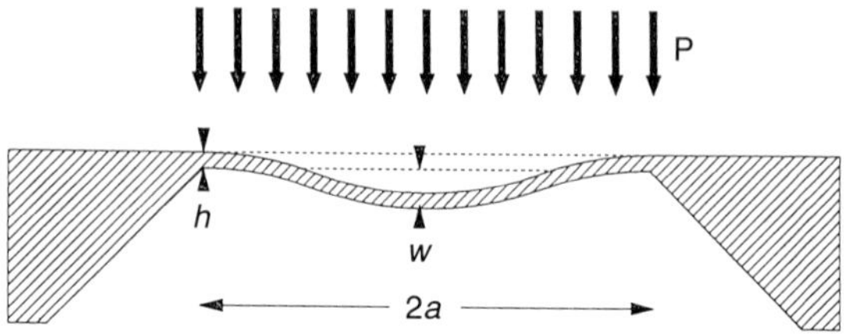

Figure 1.29. *Structure of the bulk-micromachined pressure sensor.*

to match a prescribed measurement range irrespective of membrane thickness. The deflection is either directly measured using capacitive techniques or the deflection-induced stress is measured using integrated piezoresistors. The membrane thickness is defined by the thickness of the epitaxial layer, $h = t_{\mathrm{epi}}$, which should be sufficient to avoid the pressure-induced stress from exceeding the yielding stress. Usually, more stringent conditions are imposed and a maximum stress less than 2–5 % of the yield limit stress is specified to maintain linearity. Piezoresistors of reasonable gauge factor can be fabricated using the doped layers already present in a bipolar process. Care should be taken to minimize temperature effects [1.23]. Differential pressure sensors do, therefore, not require major modifications in a standard bipolar process.

However, a few complications arise in practical devices. Firstly, the pressure sensor of most interest is the absolute type, which requires a reference cavity. The bulk-micromachined wafer can for that purpose be evacuated and hermetically sealed during packaging. Moreover, special gas access holes have to be drilled in the package. These manufacturing steps apply for any type of pressure sensor; however, in a smart pressure sensor the co-integrated circuits should be able to withstand the die bonding techniques used. This fabrication problem is, therefore, not fundamental, but does affect the economic feasibility of integrated smart pressure sensors, as will be discussed in chapter 4.

The key element in a bulk-micromachined accelerometer is the suspended seismic mass shown schematically in Fig. 1.30 [1.24]. This structure forms a mass-spring system and two additional capping wafers are used to realize a cavity that determines the

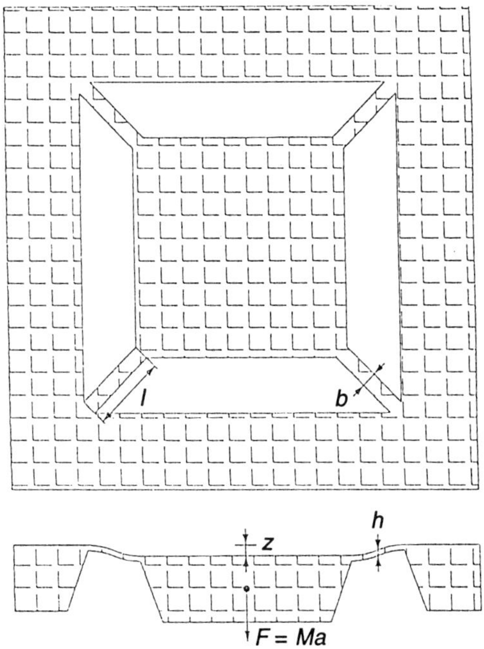

Figure 1.30. *Structure of the multi-wafer accelerometer.*

squeezed-film air damping. The sensitivity, S, and the resonance frequency, ω_o, can be calculated as:

$$S = \frac{z}{a} = \frac{Ml^3}{4Ebh^3}, \qquad \omega_o = \sqrt{\frac{4Ebh^3}{Ml^3}} = \sqrt{\frac{1}{S}} \tag{1.18}$$

respectively, where M denotes the mass, l the length, b the width and h the height of the suspension beams, as indicated in Fig. 1.30 and E the modulus of elasticity of silicon. The epilayer thickness determines the thickness of the suspension beams; $h = t_{epi}$. However, the relation clearly indicates that h is not the only possible variable to set the parameter of interest, so any sensitivity can be selected at any value of h within the limits of the yield stress by adjusting l and b. In a similar way to the pressure sensor,

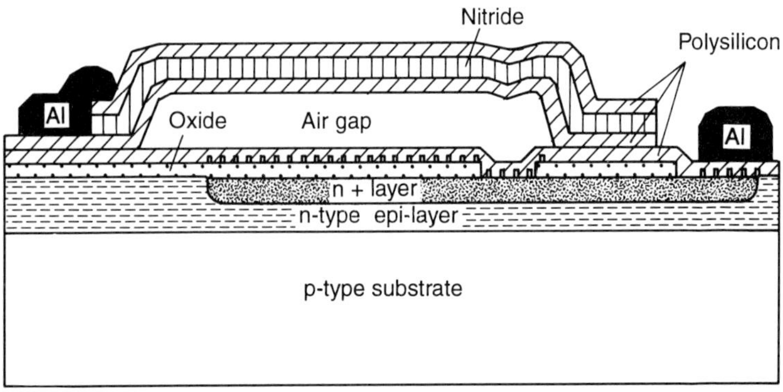

Figure 1.31. *Fabry–Perot interferometer.*

the electric constraints on the epilayer, therefore, do not imply any mechanical limitation. Fabrication of the mass wafer does not impose a major compatibility problem apart from the post-processing bulk micromachining. However, an accelerometer is a 3-wafer device and wafer-to-wafer bonding techniques are employed. Care should be taken not to endanger the compatibility in the bonding phase. Wafer bonding is discussed in chapter 2 and the economic aspects are considered in chapters 4 and 5.

Surface micromachining involves the deposition of thin layers on a silicon substrate and the subsequent selective removal of some of these layers (the sacrificial layers) in order to obtain free-standing structural layers. A typical application is the Fabry–Perot interferometer shown in Fig. 1.31 [1.25]. A nitride-poly-nitride layer is used as the structural layer. Such a multilayer sandwich makes it possible to satisfy both the optical and mechanical requirements. Integration of the photodiode and photocurrent readout circuits implies fabrication compatibility.

Fabrication compatibility with standard microelectronic processing is, in principle, maintained when the surface micromachining is carried out near the end of the standard processing. However, a number of resulting complications have to be taken into ac-

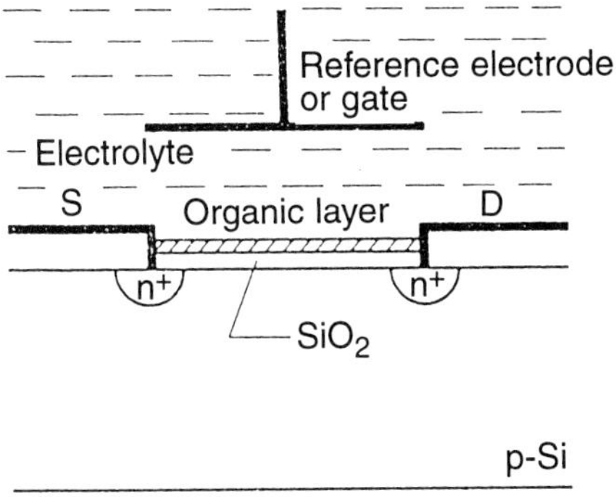

Figure 1.32. *ISFET structure.*

count. Firstly, a high-temperature residual stress anneal is required in conventionally-deposited polysilicon layers after release of the structure. Secondly, as HF is used to etch the sacrificial layer, conventional oxides cannot be used to passivate the surface during standard bipolar processing. Finally, the interconnect is evaporated or sputtered on the wafer after microstructure formation, so special care should be taken to avoid damage or sideways filling of the cavities, causing the released structure to be stuck after interconnect patterning. These issues will be discussed in more detail in the next chapter.

1.5.5 *Chemical sensors*

The most popular chemical sensor in silicon is the Ion Sensitive Field Effect Transistor (ISFET), which has a structure that is in principle directly compatible with an integrated MOS transistor with the gate (reference electrode) some distance away and the ionic solution in between. The operation of the ISFET shown in Fig. 1.32 is based on the modulation of the gate potential by the electrochemical potential of the fluid that is in contact with the channel and thus by the pH value of that fluid [1.26]. The main problem of the ISFET as a basic transducer is its limited long-term stability. However, that issue is beyond the scope of this book. The fabrication compatibility is mainly impaired at the packaging stage. The fundamental discrepancy between the need for contact between the sensor and the measurand and the protection of the circuits, as provided by the package, is most severe in chemical sensors. Direct contact is required; however, many substances of interest have a corrosive effect on the sensor surface. This dilemma has hampered developments in integrated silicon smart chemical sensors. One solution to this problem is the use of Surface Acoustic Waves (SAW) devices in which the sensing element interacts with the measurand through a chemical interface. However, piezoelectricity is required for the propagation of SAW. As silicon is not piezoelectric, compatible materials are used. ZnO will be discussed in more detail in chapter 3 as a compatible material which is able to add this capability.

1.6 Conclusions

Although the advantages initially claimed for silicon sensors should be taken with care, these are, in principle, correct. The arguments do not apply in all possible applications due to unforseen technical and economic constraints. It has been shown that the mere material compatibility of silicon with microelectronic circuits is an insufficient condition for on-chip integration and fabrication compatibility is required. The latter is far less trivial and increases the development time for a silicon smart sensor. Moreover, the small size is mainly of importance in biomedical applications and a nuisance in many others. Finally, the transduction effects available do not make silicon an outstanding sensor material. Nevertheless, the advantages of on-chip integration are unsurpassed and the concept is sound in many demanding applications, which make the research and business in silicon sensors worthwhile.

Standard bipolar and CMOS processes are reasonably tolerant to sensor processing-induced modifications. The major fabrication compatibility infringement generally results from high-temperature post processing.

Integrated silicon smart sensors are technologically feasible in a standard microelectronic process for sensing of many different parameters in the thermal, magnetic, radiant, mechanical and chemical domains. The fabrication compatibility problem is virtually non-existent in some sensor types, such as the temperature sensor, as the sensor processing gives no reason for process adjustment. Many sensors, such as the magnetotransistor, give rise to minor process adjustments and others, such as the basic color sensor, require careful circuit design.

1.7 References

[1.1] S. Middelhoek, Quo vadis silicon sensors, *Sensors and Actuators*, A41–42 (1994) 1–8.

[1.2] R.F. Wolffenbuttel, Fabrication compatibility of integrated silicon smart physical sensors, *Sensors and Actuators*, A41–42 (1994) 11–28.

[1.3] Y. de Coulon, T. Smith, J. Hermann, M. Chevroulet and F. Rudolf, Design and test of a precision servo accelerometer with digital output, *Proc. Transducer 93*, Yokohama, Japan, June 7–10 1993, pp. 832–835.

[1.4] P.M. Sarro, Sensor technology strategy in silicon, *Sensors and Actuators*, A31 (1990) 138–143.

[1.5] S. Middelhoek and S.A. Audet, *Silicon sensors*, Academic Press, London, 1989.

[1.6] S. Wolf and R.N. Tauber, *Silicon processing for the VLSI era*, Vol 1: Process technology, Lattice Press, CA, USA.

[1.7] L.K. Nanver, E.J.G. Goudena and H.W. van Zeijl, DIMES01, a baseline BIFET process for smart sensor experimentation, *Sensors and Actuators*, A36 (1993)139–147.

[1.8] S. Wolf, *Silicon processing for the VLSI era*, Vol 2: Process integration, Lattice Press, CA, USA.

[1.9] K.M. Mahmoud and R.F. Wolffenbuttel, Compatibility between bipolar readout electronics and microstructures in silicon, *Sensors and Actuators*, A31 (1992) 188–199.

[1.10] J.R. Hauser and P.M. Dunbar, Performance limitations of silicon solar cells, *IEEE Transactions on ED*, ED 24, No 4 (1977) 305–321.

[1.11] S.M. Sze, *Physics of semiconductor devices*, Wiley, 1981.

[1.12] M.I. Elmasry, *Digital bipolar integrated circuits*, Wiley, 1983.

[1.13] A.S. Grove, *Physics and technology of semiconductor devices*, Wiley, 1967.

[1.14] G.C.M. Meijer, Thermal sensors based on transistors, *Sensors and Actuators*, 10 (1986) 103–125.

[1.15] A.W. van Herwaarden and P.M. Sarro, Thermal sensors based on the Seebeck effect, *Sensors and Actuators*, 10 (1986) 321–346.

[1.16] R.S. Popović, *Hall-effect devices*, Hilger, Bristol, UK 1991.

[1.17] R.S. Popović and H.P. Baltes, A CMOS magnetic field sensor, *IEEE J. Solid-State Circuits*, SC 18 (1983) 426–428.

[1.18] B. Hälg and R.S. Popović, How to liberate integrated sensors from encapsulation stress, *Sensors and Actuators*, A21–A23 (1990) 908–910.

[1.19] C. Riccobene, G. Wachutka, J. Bürgler and H. Baltes, Two-dimensional numerical modelling of dual-collector magnetotransistors: evidence for emitter efficiency modulation, *Sensors and Actuators*, A31 (1992) 210–214.

[1.20] R.F. Wolffenbuttel and G. de Graaf, Performance of an integrated silicon colour sensors in terms of response to colours in the colour triangle, *Sensors and Actuators*, A21–A23 (1990) 574–580.

[1.21] R.F. Wolffenbuttel, Photodiodes in silicon with an electronically-programmable UV response, *Sensors and Actuators*, A21–A23 (1990) 559–563.

[1.22] J.A. Kemmer, Silicon detectors for nuclear radiation, *Sensors and Actuators*, 15 (1988) 169–184.

[1.23] R.F. Wolffenbuttel, Integrated micromechanical sensors and actuators in silicon, *Mechatronics* Vol. 1 No 4, 1991, pp. 371–391.

[1.24] F. Rudolf, A. Jornod, J. Bergqvist and H. Leuthold, Precision accelerometers with μg resolution, *Sensors and Actuators*, A21–A23 (1990) 297–302.

[1.25] K. Aratani, P.J. French, P.M. Sarro, D. Poenar, R.F. Wolffenbuttel and S. Middelhoek, Surface micromachined tunable interferometer array, *Sensors and Actuators*, A43 (1994) 17–23.

[1.26] P. Bergveld, The development and application of FET-based biosensors, *Biosensors*, 2 (1985) 65–118.

2
DEVELOPMENT OF COMPATIBLE MICROMACHINING PROCESSES IN SILICON

2.1 Introduction

When combining micromachining devices with electronics a number of problems may arise. The primary problems are the restriction on the thermal budget due to the diffusions in the wafer and compatibility of etchant with both the processed wafers and the clean room environment. This may present considerable restrictions on the micromachining processing.

Micromachining can be divided into two main groups: *bulk*, where structures are etched in the substrate; and *surface*, where the micromechanical layers are formed from layers deposited on the surface. In the case of bulk micromachining, compatibility problems generally arise from the incompatibility of the processing with the clean room environment and this can often be solved by performing the micromachining at the end of the processing. In the case of surface micromachining there are several considerations. The first is the thermal budget, since the micromechanical layers may require annealing. The second is non-planarity of the wafer surface caused by the deposition of multiple layers. The last major consideration is

Table 2.1 Types of mechanical structures

Static	Nozzles, cavities, capillary columns, electrical connectors
Dynamic	Membranes, microbridges, cantilevers, resonators
Kinematic	Micromotors, microgears, pin joints, springs, cranks, sliders

Table 2.2 Properties of the most important materials in micromachining

	Melting point (°C)	Thermal expansion ($\times 10^{-6}$°C^{-1})	Density (gcm^{-3})	Young's modulus (10^{11}Pa)	Yield strength (10^{9}Pa)
Si	1415	2.5	2.4	1.3–1.69	6.9
SiN	1900	2.8	1.48	2.43	14.0
SiO_2	1610	0.5	2.27	0.73	8.4
Al	660	25	2.70	0.70	0.17

the etchant required to remove the sacrificial layer. In many cases this etchant will also attack the electronic components.

The term micromechanical includes any mechanical structure fabricated in or on the silicon wafer. Micromechanical structures can be divided into three groups: *static*, *dynamic* and *kinematic*. Examples of these structures are shown in Table 2.1 [2.1].

For all these structures the mechanical properties of the materials are an important consideration. These properties are listed in Table 2.2, for commonly used materials.

In traditional silicon sensors the sensing element yielded an analog signal which was fed via cables to the signal processing and readout electronics (Fig. 2.1a). The set-up had the limitations of noise. Furthermore there are various deviations of the characteristics from the ideal, which can create inaccuracies. These deviations include non-linearity, hysteresis, drift and cross-sensitivity.

To improve the system, multi-packaging was introduced where the sensing and signal processing chips were bonded into a single package or on a thick-film substrate, (Fig. 2.1b). However, with the demands for high accuracy it has become necessary to process

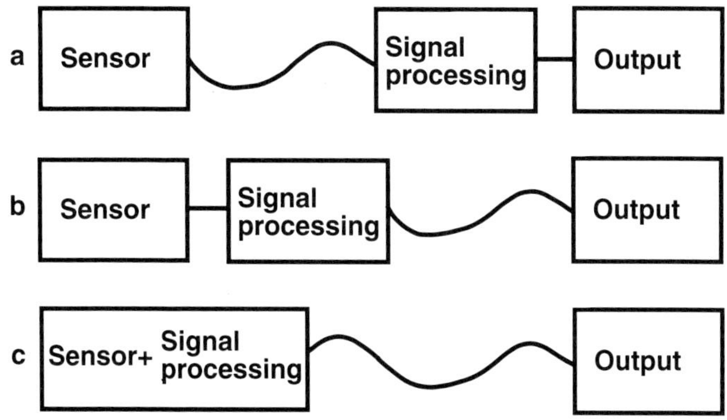

Figure 2.1. *The stages of development of smart sensors: (a) completely separated sensor and signal processing, (b) hybrid solution and (c) smart sensor.*

the sensor signal as close as possible to the measurement point. The small size of micromachined mechanical devices results in extremely small electrical output signals and therefore the need arises for smart micromachined sensors (Fig. 2.1c). If, for example, a capacitance change is in the order of fF in a capacitance of a pF, it is essential that the signal is processed without the use of bonding wires. With the smart device the signal can be compensated for temperature, offset or non-linearity and converted into a digital signal, which is less prone to noise, on the sensing chip. This solution has the disadvantage that the processing of the sensor and electronics must be compatible. This chapter discusses the development of micromachining processes compatible with electronics and the problems which can arise.

Once it has been established that the micromachined structures are to be integrated with the electronics, it is necessary to establish the process limitations and whether additions to the standard process are to be permitted. The devices can be integrated with CMOS [2.2–2.3] or bipolar [2.4] processes. It is important to establish a strategy in terms of how the additional steps required for the micromachining can be inserted into the standard process. Such a discussion is given in [2.5] by Sarro.

With bulk micromachining sections of the silicon substrate are removed to produce membranes, beams and mesas. This processing is often performed after the completion of the electronic processing as, for example, KOH etching, which is frequently used, is not considered compatible with a clean room environment. Surface micromachining, on the other hand, involves the deposition and selective removal of additional layers on the silicon surface. The important question with surface micromachining is when this additional processing should be performed to ensure that both the electronics and the micromechanics can be optimized.

One of the advantages of surface micromachining on silicon is that, due to its inherent small size, it is possible to include more electronics on the same chip. To achieve this, a micromachining process has to be combined with a CMOS or bipolar process, without influencing the characteristics of the electronic devices. Many compatibility problems must therefore be solved, before the process is ready to be used by the designers. A major consideration in the development of a process is the thermal budget, as any additional high-temperature step can have a considerable influence on the bipolar characteristics [2.6]. A bipolar process is able to withstand limited additional thermal processing, such as at 800–900°C. If higher temperatures are used after the formation of the base and emitter, the increased diffusion can seriously affect the transistor characteristics. Many of the bulk micromachining processes, on the other hand, involve back-side etching with KOH or EDP and do not therefore present problems of thermal budget.

It is essential to establish a process where the characteristics of both the mechanical and electrical components can be maintained. Furthermore, the designer must know the limitations in terms of the additional steps which can be added, and both these considerations are discussed below.

2.2 Compatibility considerations

An important consideration when developing a sensor with in-built signal processing is first to establish from where the compatibility problems may arise.

The considerations when integrating the sensor can be divided into two groups: 1) compatibility of the process with the clean room

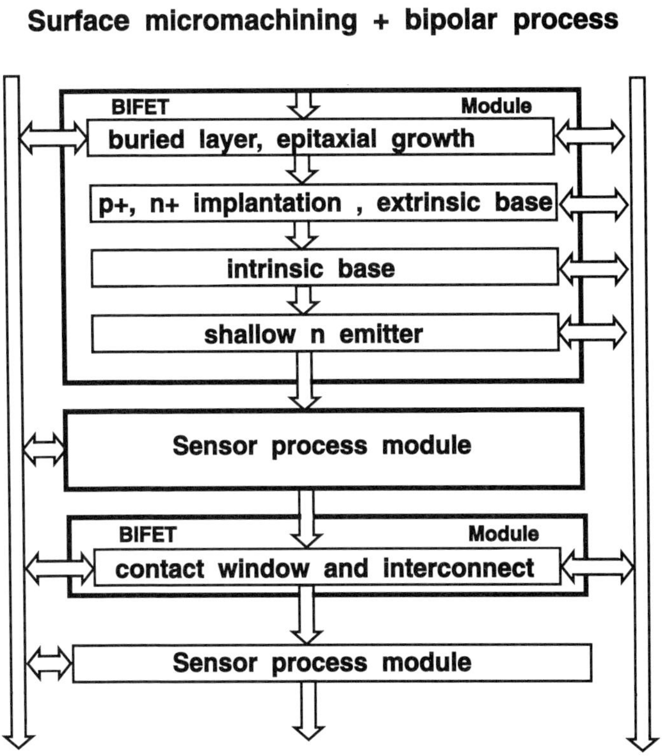

Figure 2.2. *Bipolar processing sequence with inserted sensor modules.*

and, 2) compatibility between the sensor and circuits. In the case where the process is not compatible with a clean room then this can only be used if the processing step is performed after the clean room process has been completed. This is the solution normally applied to KOH etching in bulk micromachining. The second problem is that of the compatibility of the micromachining processing with that of the electronics.

Whether the circuitry is fabricated in a bipolar or a CMOS process, it is essential to establish at what stage the additional processing can be inserted. The example of a bipolar process scheme is shown in Fig. 2.2 [2.4]. Within this process there are two viable

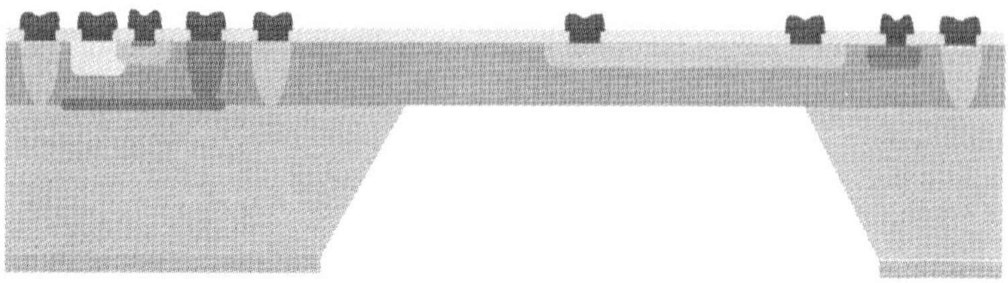

Figure 2.3. *Typical bulk micromachined structure.*

options. The first is to insert the micromachining processing before the emitter anneal and to use this anneal for both the emitter and the micromachined layers. The second option is to add the micromachining phase after the completion of the bipolar process, but before contact opening and aluminum deposition. In this case, the total thermal budget available for the micromachining is low as the emitter junction depth is extremely sensitive to any additional steps. In both cases it must be established whether the sacrificial etching is performed before or after aluminum formation.

2.3 Bulk micromachining — back-side

Bulk micromachining is the process of etching wells in the silicon substrate leaving, for example, membranes, mesas or beams. Silicon has excellent mechanical properties and lends itself well to micromachining. The mechanical properties of the material are discussed in detail by Petersen [2.7]. Accurate crystal orientation-dependent etching, presented by Bean [2.8], has led to a major break-through in the use of silicon as a micromachined material.

Anisotropic wet etchants use the fact that some crystal orientations etch much more slowly than others. For example using a (100) wafer will yield the etch cavity shown in Fig. 2.3. In this case the etching has stopped on the <111> plane. This effect is due to the energy required to remove an atom from the surface and the atom spacing revealed to the surface. The large differences between etch-rates of silicon planes can present problems when etching corners and this will be discussed below.

Much of the traditional bulk micromachining involved wet etching from the back-side of the wafer, stopping on the epitaxial layer. The etch-stop often uses electrochemical techniques and therefore a fully processed wafer is used where contact to the aluminum is made via a clamp in the wafer holder. In terms of design it must be ensured that the aluminum has relatively large contacts close to all the areas to be etched to ensure uniform etching. Bulk micromachining can also be performed from the front-side and this will be discussed in a later section.

Bulk micromachined structures have been used for a wide variety of structures for many years. These include thermal of flow devices where thin membranes are formed to improve heat dissipation [2.9–2.10], pressure sensors [2.11], accelerometers where a micromachined mesa is supported by thin beams of epitaxial thickness [2.12–2.13], micro-pumps [2.14–2.15] and optical devices [2.16]. In some cases the micromachined structures have been combined with relatively complex electronics in a feedback system [2.17]. An example of an accelerometer is shown in Fig. 2.4 [2.13]. All the above mentioned structures require several additional steps after the completion of the electronics processing. These include the etching described in this section and also, in many cases, some form of silicon or glass bonding; this will be dealt with in a later section.

The basic electrochemical etching set-up is shown in Fig. 2.5. In the case of KOH etching, the masking material is usually nitride. The etching of silicon does not require a current flow from the platinum electrode. This is however required for the etch-stop. The epilayer is biased as the anode and the platinum as the cathode. When the sample is immersed in the etchant, the etching begins. In the case of an electrochemical etch-stop, the electrical potential at the etch plane when the epitaxial layer is reached ensures that the etching does not proceed. By further consideration of the open-circuit-potential (OCP) and passivation potential (PP) [2.18] a greater degree of flexibility can be achieved. If the epitaxial biasing is greater than 0.6 V, the epitaxial layer is passivated and the etching is stopped. If the voltage drops below this value, the etching will be re-started. Thus the electrochemical etchstop

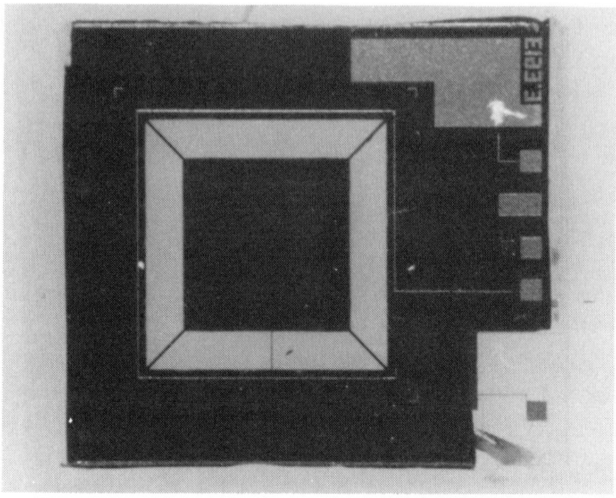

Figure 2.4. *Example of a bulk micromachined accelerometer.*

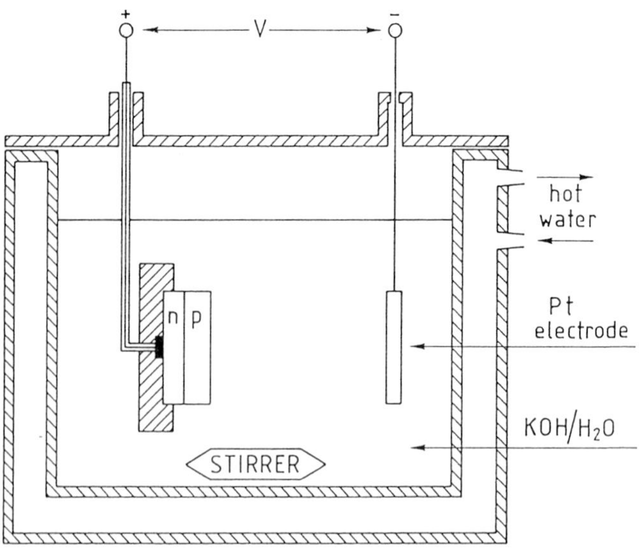

Figure 2.5. *Electrochemical etching set-up.*

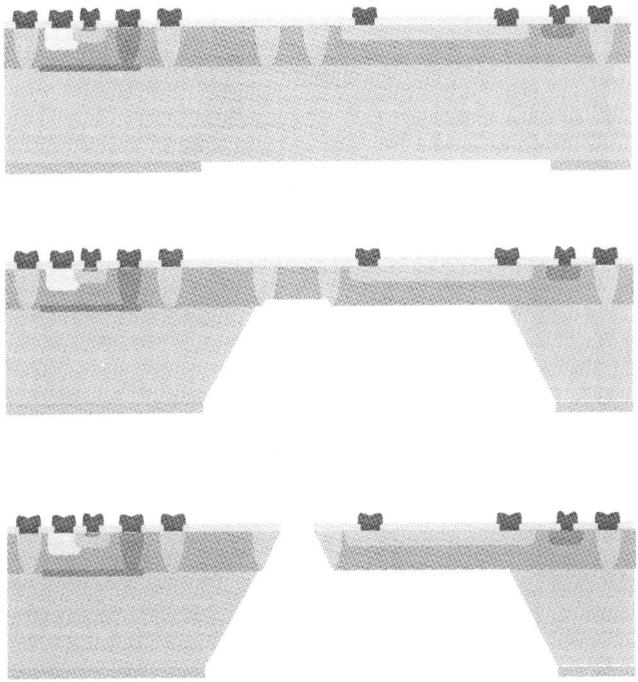

Figure 2.6. *Processing sequence for bulk micromachined device where the final etch through the epi is performed using back-side KOH etching.*

provides a highly reliable method to control the thickness of a beam or membrane. The etchstop mechanisms for various etchants is discussed below. The mask on the back-side is of silicon nitride and requires special alignment techniques. Two possible techniques for easing this problem are: (1) the forming of special alignment marks at the beginning of the process [2.19] and (2) forming a potential pattern in the substrate, via electrodes on the silicon surface [2.20].

In many cases it is necessary to etch through the epitaxial layer in selected areas and this can be achieved by two methods. The first is to ensure certain epitaxial layers are not passivated by isolating them from the main epitaxial areas. The second is to etch from the front-side. These two techniques are shown in Figs 2.6 and 2.7.

There are several suitable etchants and each must be fully char-

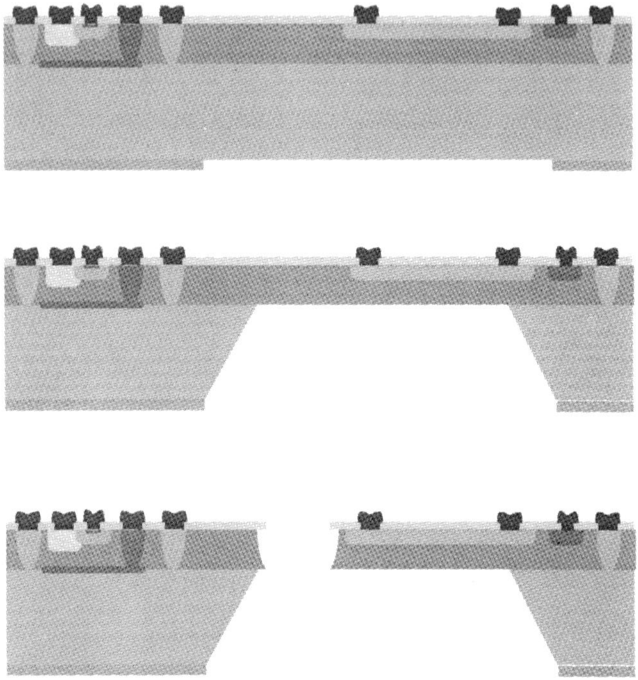

Figure 2.7. *Processing sequence for bulk micromachined device where the final etch through the epi is performed using front-side etching.*

acterized in terms of etch rate, masking materials and compatibility with circuit processing/clean rooms. Etchants commonly used are hydrazine-water, EDP (ethylenediamine-pyrocatechol-water), KOH (potassium hydroxide) and TMAH (tetramethylammonium hydroxide). The main characteristics of these etchants are listed in Table 2.3 [2.1,2.7,2.21]. All characteristics of the etchant are dependent upon the concentration and temperature and only commonly used examples are quoted here.

The etchant KOH is probably the most commonly used etchant. One disadvantage is the relatively high etch rate of silicon dioxide. Silicon nitride is therefore a commonly used masking layer. If PECVD nitride is used, the whole bulk micromachining processing can be performed after the electronic processing has been com-

Table 2.3 Characteristics of major wet etchants for bulk micromachining

Etchant	Suitable Masking		Etch rate		Comments
		(100)/ (111)	SiO_2 Åhr^{-1}	SiN Åhr^{-1}	
Hydrazine	SiO_2, SiN, Most metals	16:1	100	$\ll 100$	Highly toxic. Explosive at high concs.
EDP	Au,Cr,Ag, Au,Ta, SiO_2, SiN,	35:1	120	60	highly toxic
KOH (IPA) 30–40wt% 60–90°C	SiN, Cr, Au	up to 400	1700–3600	<1	Not clean-room compatible
TMAH +IPA	SiO_2, SiN,	max. 300	< 100	< 20	

pleted. In some cases the etching properties have been enhanced by the introduction of isopropyl alcohol (IPA) [2.22].

An alternative etch-stop available with some etchants is a high boron etch-stop ($> 2.5 \cdot 10^{19} cm^{-3}$). If the boron etch-stop is used, EDP has an extremely sharp cut-off, and is more suitable than KOH as shown in Fig. 2.8 [2.23]. However, a standard bipolar process does not have a suitable buried layer, and therefore would have to be added. In this case, particular care must be taken to ensure that such a highly doped layer does not affect other devices.

All the etchants operate at low temperature and can be masked using low-temperature deposited layers. They are all therefore compatible with the electronics in terms of thermal budget. Thus, provided a suitable etch-stop can be incorporated into the design, a smart sensor device can be fabricated. Such structures have been incorporated in both bipolar [2.13] and CMOS [2.2] processes. Both use <100> silicon wafers and many modern CMOS processes use p and n wells in an n-type epi. A further example of the flexibility of these processes is the combination of bulk micromachining, electronics and zinc oxide in the fabrication of surface acoustic wave devices [2.24].

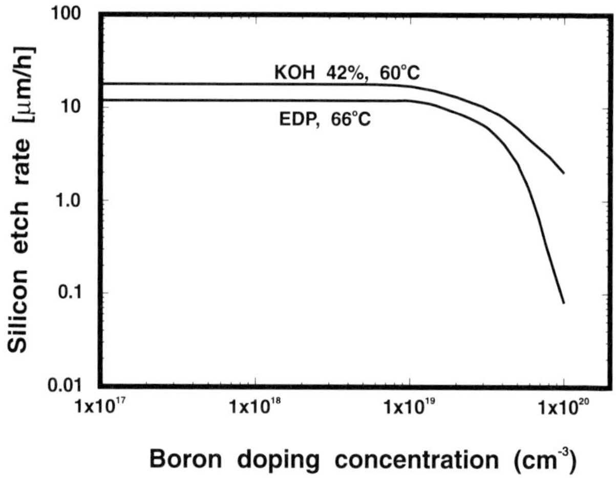

Figure 2.8. *Etch rate as a function of boron doping concentration for both KOH and EDP. Taken from Seidel et al. [2.23].*

An important design consideration is corner compensation. When the etchant meets a convex corner the plane revealed may have a high etch rate, resulting in the removal of the entire corner [2.25–2.26], and for a smaller structure the possible removal of the whole mesa. Various designs have been developed and an example of these is shown in Fig. 2.9 [2.26]. The principle of the method is to add thin beams or squares to the corners which are rapidly under-etched. If the extension is large enough, the desired corner will be reached just as the etching reaches the epitaxial layer. If the etching is continued, the corner will be removed as before. This etching of the corner compensation structure is shown in Fig. 2.10. Perfect compensation cannot be achieved and for some structures it has been shown that ridges remain on the membrane, as shown in Fig. 2.11 [2.26].

When considering the most suitable etchant for compatible bulk micromachining, the thermal budget is not a consideration. Table 2.3 lists 4 examples of possible etchants. The first two etchants, hydrazine and EDP, have the disadvantages of being highly toxic and therefore difficult to handle. The most commonly used etchant

Figure 2.9. *Corner compensation structure.*

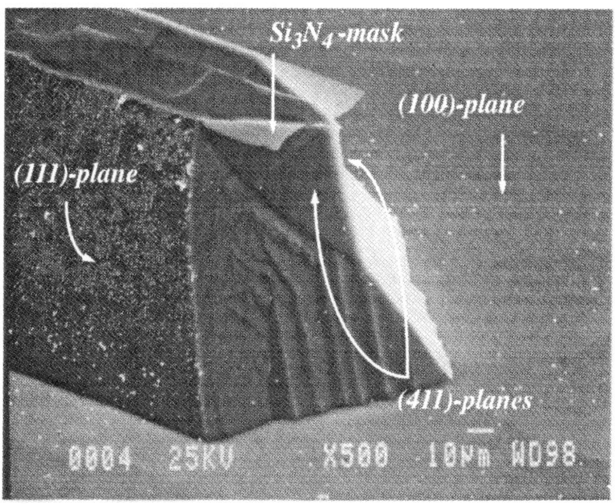

Figure 2.10. *Under-etching of a corner compensation structure.*

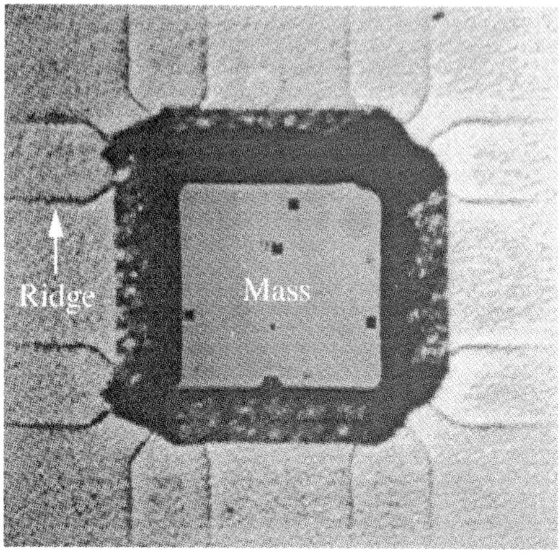

Figure 2.11. *Ridges remaining on the membrane from corner compensation structures.*

is KOH. This is relatively easy to handle but the potassium deposits of the silicon surface are not compatible with a clean room environment. The etching must therefore be performed in a separate area and the wafers should not be returned to the clean room. Recently, there has been increasing interest in TMAH an alternative. The wafers can be cleaned more easily after etching and can therefore be returned to the clean room for further processing if necessary.

2.4 Bulk micromachining — front-side

There are two limiting factors with back-side bulk micromachining: (1) the dimensions are usually large and (2) alignment is difficult and has limited accuracy (although recent improvements in back-side mask alignment techniques have greatly reduced this problem). If the bulk micromachining can be performed from the front-side of the wafer, this alignment problem is automatically removed and the dimensions can be reduced.

Membranes and beams can be produced on the surface by applying the wet etchant, such as KOH described above, to the front-side of the wafer. First the front-side must be protected with a nitride layer. In this case a PECVD layer must be used as it is needed to protect the aluminum and therefore the temperature must be kept low. The bulk micromachining steps can be added to the end of the standard process. Use can be made of layers deposited or grown on the surface of the wafer, where the sacrificial layer is the underlying silicon [2.2,2.27]. Alternatively, both the mechanical and the sacrificial material can be silicon. Both these techniques have been applied to structures such as flow sensors and resonators. In recent years there has been increased interest in the combination of silicon microelectronics and optics. Front-side bulk micromachining has be used to fabricate channels for fibers [2.28]. By careful consideration of the orientation dependence of the etching, vertical mirrors and beam splitters can be fabricated [2.29]. Combinations of wet and dry etching on the front-side of the wafer have also been applied to the fabrication of the micro-tip [2.30–2.31].

If deep trenches are to be fabricated it is preferable to perform the etch at the end of the process (i.e. after aluminum deposition). This can be done using plasma etching for which two techniques have been proposed recently. The first, known as SCREAM, uses a combination of anisotropic and isotropic plasma etching [2.32]. The basic process sequence is shown in Fig. 2.12. First trenches are etched which define the sidewalls of the structure. The sidewalls are protected by an oxide deposition and etch-back using plasma etching. A further plasma etch etches the trench beyond the sidewall protection and finally the structure is under-etched using an isotropic etch. However, this process uses a trench surrounding each micromechanical element, followed by a final aluminum deposition, to make electrical contact. These final two steps make true integration with electronics extremely difficult.

The second technique, know as SIMPLE, can be achieved using a single etch [2.33]. This process makes use of a plasma which etches low-doped material anisotropically and n-type material above a threshold of about $1 \cdot 10^{20}\text{cm}^{-3}$. The basic process sequence is shown in Fig. 2.13. This technique has been used to fabricate micromachined devices and a cross-sectional SEM photograph of the final structure is shown in Fig. 2.14. A high aspect ratio can

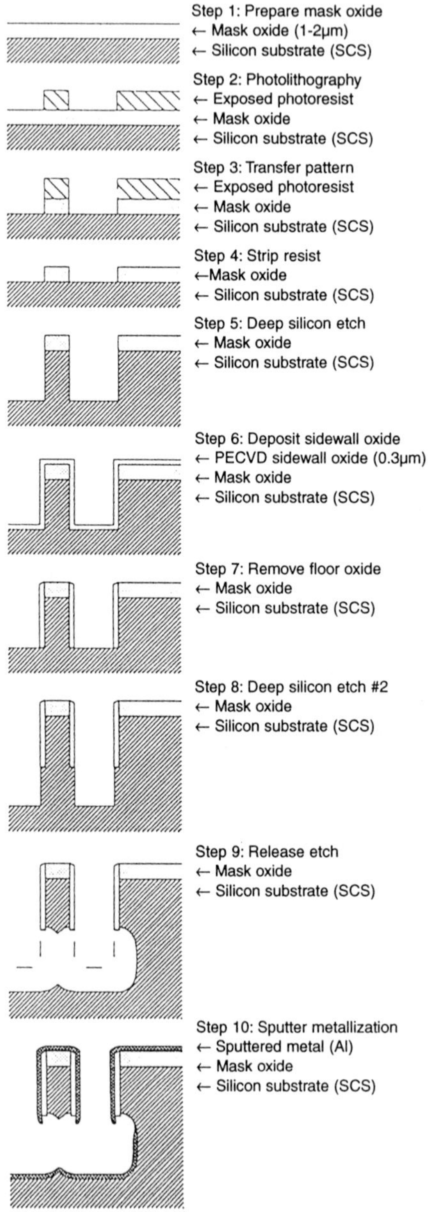

Figure 2.12. *SCREAM process sequence. Taken from Shaw et al. [2.32].*

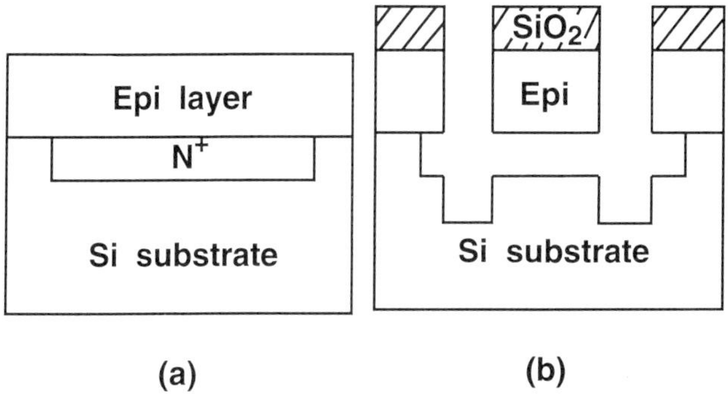

Figure 2.13. *Basic processing sequence for the SIMPLE process; (a) vertical trench etching, (b) under-etching.*

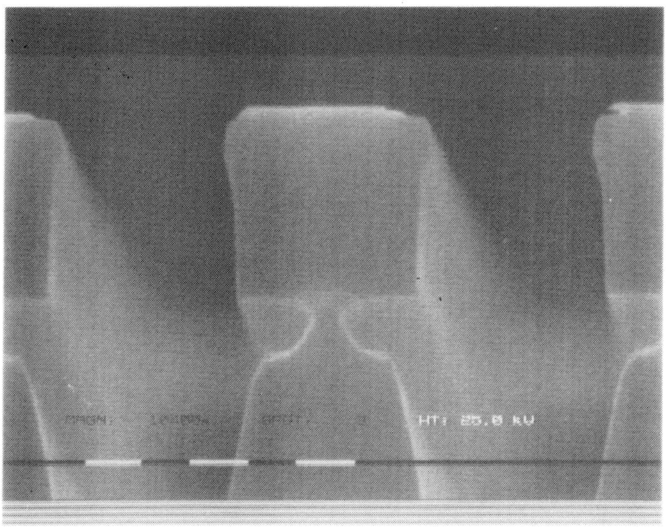

Figure 2.14. *SEM photograph of structure etched using the SIMPLE process [2.33].*

be achieved, where the height of the device is determined by the epitaxial thickness. The width of the structure shown here is 2 μm. The second technique has the advantage of simplicity. This process is performed after the aluminum deposition and is fully compatible with the electronic circuitry.

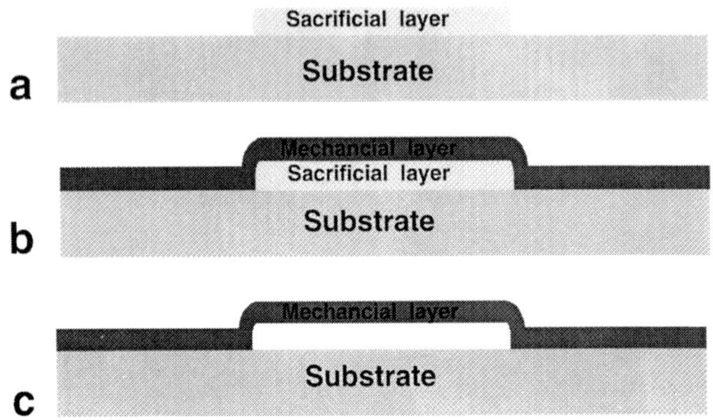

Figure 2.15. *Basic surface micromachining process: (a) deposition and patterning of the sacrificial layer, (b) deposition and patterning of the micromechanical layer, (c) sacrificial etching.*

2.5 Surface micromachining

With bulk micromachining, areas of the silicon substrate are etched to form membranes and mesas. In microelectronics terms these structures are usually relatively large. Surface micromachining, on the other hand, is based on the deposition of thin films on the surface of the wafer and removing one or more layers to release the mechanical structures. A simple single layer surface micromachining process is shown in Fig. 2.15. All processing steps up to sacrificial etching are standard deposition and etching steps. The sacrificial layer is first deposited and patterned, as shown in Fig. 2.15a. The desired mechanical layer is deposited, annealed (if necessary) and patterned (Fig. 2.15b). The final step is the sacrificial etching, where the layer is etched through holes leaving a free standing structure, (Fig. 2.15c). The additional steps, such as contact windows and metallization will be discussed in a later section.

There is a wide range of both sacrificial and mechanical layers available to the designer, where the sacrificial layer depends upon the required mechanical layer. Some of these layers, such as polysilicon and nitride, can be incorporated into a standard silicon process. Other layers, such as tungsten, molybdenum, polyamide, TiNi or ZnO, must be added to the end of the process. This section will concentrate on the two layers which can be incorporated into the standard process. For these two materials the use of an oxide sacrificial layer will be discussed.

Initially a decision has to be made whether the surface micromachining should precede or follow the thermal processing of the electronics. If the surface micromachining is done before the electronics, this has the advantage that the characteristics of the electronic devices are not degenerated by any further processing. However, the mechanical characteristics of surface micromachined structures may suffer from the many high-temperature steps necessary in bipolar processing. Furthermore, the poor planarity of the wafer after processing of the surface micromachined structures, may cause problems with defining small structures, due to inhomogeneity of the resist layer.

The requirements for good compatibility with standard electronics implies that the surface micromachining process must either follow the electronic processing, or be inserted at a late stage, where the last annealing step is suitable for both structures.

In this section the development of a surface micromachining compatible with bipolar electronics will be described. The sacrificial layer described is phosphor-silicate glass (PSG) and the mechanical layers, polysilicon and nitride. All three are deposited layers using LPCVD and have a deposition temperature compatible with the electronics, provided the deposition is performed before aluminum deposition. Firstly the optimization of the deposited layers will be discussed, followed by some of the practical problems arising when they are combined with electronics. This will lead to the development of a fully compatible surface micromachined process [2.34].

2.5.1 The sacrificial layer

A surface micromachining process requires a sacrificial layer which is removed at a later stage to release the micromechanical parts. Many materials have been quoted in the literature, such as: porous silicon [2.35], aluminum [2.36], polysilicon [2.37] and oxide (using SOI [2.38], SIMOX [2.39], LOCOS [2.40] and deposited oxide [2.41]), each requiring a specific etchant. As mentioned above, this section will concentrate on oxide sacrificial layers which are removed by wet etching in HF. There are various types of oxide, all of which have certain advantages and disadvantages. These options are listed in Table 2.4.

All these layers have been used as sacrificial layers. PSG with its high etch rate and low deposition temperature has distinct advantages. The high etch rate of the PSG is mainly due to the presence of phosphorus, but one potential problem is structural modification during subsequent thermal processing [2.31]. The main effects are: shrinkage and etch rate reduction.

Typical deposition parameters for LPCVD PSG layers are as follows [2.32]:

- temperature: 425°C
- pressure: 200 mTorr
- oxygen flow: 200 sccm (standard cubic centimetre)
- phosphine flow: 2–10 sccm.
- silane flow: 46 sccm
- deposition rate 50–100 $Å min^{-1}$.

The considerable increase of both etch rate and under-etch rate with increasing phosphorus doping is shown in Fig. 2.16. Both vertical and under-etch rate show an exponential dependence on the phosphorus content. Vertical etch rates in 1% HF larger than 50 $Å s^{-1}$ are obtained for the as-deposited PSG for PH_3/SiH_4 flow ratios larger than 15%. After annealing (850°C/30 minutes in dry nitrogen) the etch rate reduces, although values over 50 $Å s^{-1}$ can be achieved for PH_3/SiH_4 flow ratios larger than 21%. The lateral under-etch rate in 20% HF are presented together (for comparison, the vertical etch rate of thermal oxide in 1% HF is 0.3 $Å s^{-1}$ and its under-etch rate is 0.3 $\mu m min^{-1}$).

Annealing of PSG at 850°C results in a considerable reduction in etch rate, even though there is no significant reduction in phosphorus content. Densification is also unable to explain the reduction as shown in Fig. 2.17, and changes are likely to be due to chemical bond changes.

The increased shrinkage with increasing phosphorus content, as indicated by Fig. 2.17, is to be expected as more phosphorus oxides are formed and, consequently, the structure of the PSG is more unstable and affected to a larger extent by the anneal. The variation of the shrinkage with the phosphorus content indicates that low

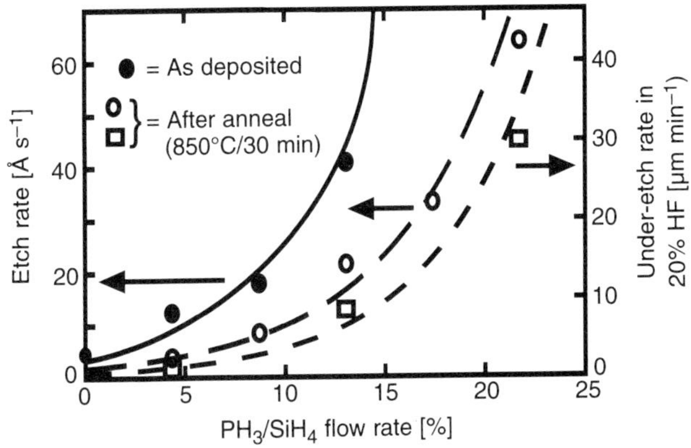

Figure 2.16. *Etch rate and under-etch rate as a function of phosphorus content in the PSG.*

phosphorus concentrations in PSG are preferable, if low shrinkage is required.

The modification induced by the anneal is extremely rapid. As shown in Fig. 2.18, the shrinkage and the vertical etch rate reach stabilized values after the first few minutes of annealing, even at an anneal temperature of 850°C. The reduction in etch rate is dependent on anneal temperature as shown in Fig. 2.19. Once annealed, however, further annealing at the same, or lower temperature have no additional effects of the PSG. Therefore provided the PSG is annealed before further processing, the film remains stable.

2.5.2 The micromechanical layers I — polysilicon

Polysilicon is a material frequently used for both electronics and sensors. In the field of electronics it has been applied to the polysilicon emitter bipolar, the polysilicon gate MOS and interconnects. Since the piezoresistive effect in polysilicon was first

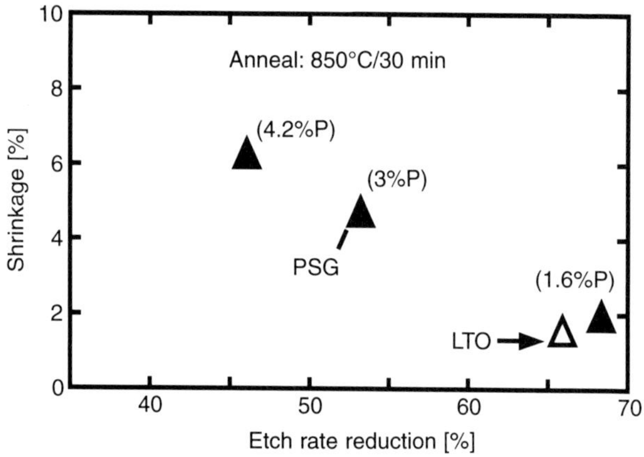

Figure 2.17. *Shrinkage versus etch rate reduction for PSG films.*

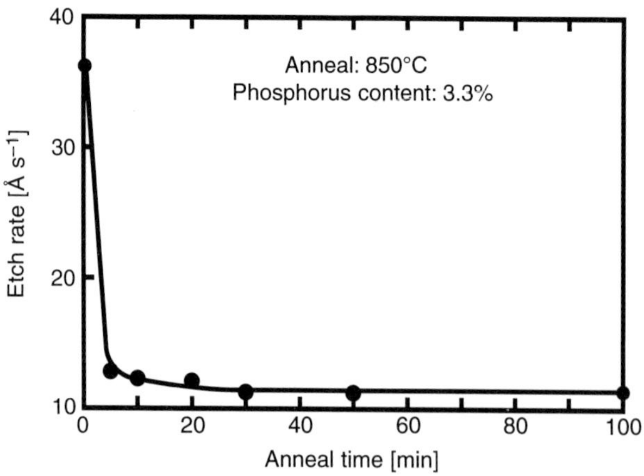

Figure 2.18. *Etch rate reduction of PSG as a function of anneal time.*

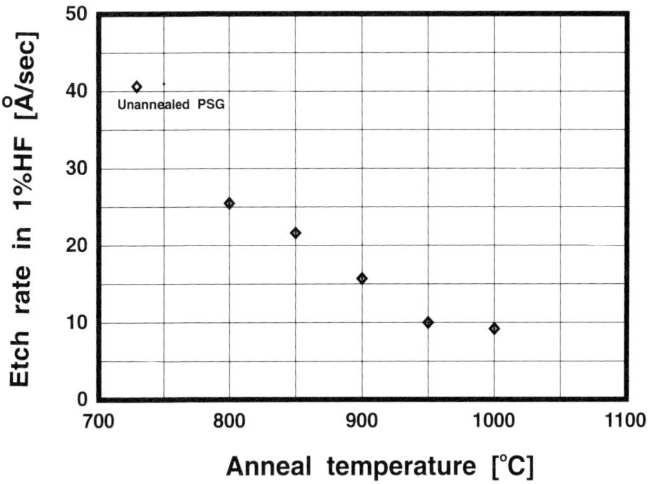

Figure 2.19. *Etch rate reduction of PSG as a function of anneal temperature.*

studied in the early 1970s [2.42], there have been many studies of its electrical [2.43–2.44] and piezoresistive properties [2.45–2.46]. With the development of surface micromachining techniques has come an increasing interest in the mechanical properties of the material [2.47–2.52].

There are two processing step after deposition which may require high temperatures. These are to activate doping (for ion-implanted polysilicon) and a stress anneal. It is therefore essential that the mechanical properties of all the micromechanical layers are characterized in order to develop a low-temperature process.

In the development of a surface micromachining process it is important to be able to characterize the layers in terms of mechanical strain. This can be achieved by the buckling technique [2.53] but this has the disadvantage of requiring an array of structures and accuracy of the measurement is dependent upon the number of structures. A second technique, used for measuring tensile strain, measures directly the displacement of the intersection of a two beams [2.54a and b]. This is, however, only useful for relatively high strain levels. A simple technique to measure both

Table 2.4 Oxide options as sacrificial layers

	Layer	Advantages	Disadvantages	Etch rate BHF
Thermal	Oxide	High quality Standard processing Good thickness control	High thermal budget Low etch rate	13Ås^{-1}
LPCVD	Low temperature oxide (LTO)	Low temperature	Low etch rate	61Ås^{-1} (unannealed)
	Phospho-silicate-glass (PSG)	Low temperature deposition High etch rate	Thickness uniformity Out-diffusion of phosphorus	Up to 600Ås^{-1} (unannealed)
	Boro-phospho-silicate-glass (BPSG)	Low temperature deposition Low reflow temperature	Thickness uniformity Low etch rate	Up to 55Ås^{-1}
PECVD	Oxide	Low temperature	Low etch rate	30Ås^{-1}

compressive and tensile strain using a pointer structure is shown in Fig. 2.20 [2.55]. Here a wide range of strains can be measured with a single structure. The thickness of the turning point, K, introduces an error, but this can be compensated using a simulated correction factor.

In this section the optimization of various processing parameters, developed at Delft University, with reference to the mechanical stress is described. For most structures a low tensile strain is required. If the layers are in compressive strain, they tend to buckle when released and can no longer be used. High tensile strain may lead to cracking and eventual structural failure of the devices. The optimal level of tensile strain depends on the structure of the device.

A method frequently used for the relaxation of residual stress in polysilicon films is a high-temperature ($> 1000°C$) anneal for an hour or more [2.50]. This yields the desired mechanical charac-

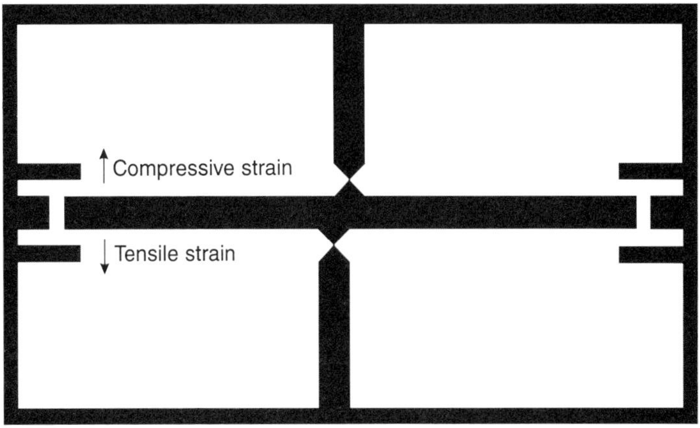

Figure 2.20. *Micromechanical structure used for measuring film strain [2.54].*

teristics but is obviously not compatible with bipolar electronics. It is furthermore advantageous to be able to use film thicknesses commonly used in standard processing.

Processing parameters have a considerable effect on the electrical and mechanical properties of polysilicon [2.46]. The important processing parameters include: deposition temperature, *in-situ* [2.48] annealing, doping and post-implant annealing. An additional parameter which should be considered is the underlying layer, which can also affect the mechanical properties of the polysilicon, even though it is removed during processing.

The polysilicon described in this section was deposited in a standard LPCVD reactor using a reactor pressure and silane flow of 150 mTorr and 45 sccm, respectively. The temperature range of deposition was 560°C–610°C. An additional option is to use an *in-situ* anneal in nitrogen, first proposed by Guckel [2.48], directly after deposition to adjust the mechanical properties of the material. Doping of the polysilicon can be achieved by deposition using $POCl_3$ or implantation. The processes described here concentrate on implantation of phosphorus and boron for a range of energies and doses.

As mentioned above, initial studies concerned the development of an optimum polysilicon process. The main criterion is the effect of strain levels, although sheet resistance is also of interest. It is important to consider each step in the process not only to optimize the process, but also to establish any flexibility for changes where may exist.

Deposition temperature is known to have a strong effect on both the electrical and mechanical properties of polysilicon. Above a certain temperature, which depends on both pressure and gas flows, the films are deposited as polysilicon with fully formed grains and grain boundaries. Below this temperature a semi-amorphous film is deposited. In this case, grains are formed but separated from each other by an amorphous layer. The transition temperature can be established from sheet resistance, since it corresponds to a peak resistance. This effect is shown for polysilicon, deposited on thermal oxide and doped with phosphorus with an implantation dose of $1 \cdot 10^{15}\mathrm{cm}^{-2}$, in Fig. 2.21. Lower doping levels have a higher sensitivity to changes in the polysilicon structure. This sharp transition is due to a change in the barrier height at the grain boundary [2.46]. A sharp transition also exists for strain, as shown in Fig. 2.22. It can be seen in Fig. 2.22 that the transition is slightly higher for strain than for sheet resistance. This is likely due the narrowing of the grain boundary causing a change in the electrical characteristics before a change in the mechanical characteristics.

By depositing as a semi-amorphous films it can be ensured that the thin grain boundary is formed by annealing and not during deposition, which reduces the stress contribution from the grain boundary. Furthermore, the resulting small grains lead to improved stress distribution, greater repeatability and smoother surface (the last factor is of particular importance where moving parts are required). It can be seen from Fig. 2.22 that a further reduction in deposition temperature does not greatly affect the strain levels. Thus, maintaining a deposition temperature as close to the transition as possible maximizes the deposition rate while maintaining the desired mechanical properties. Films deposited near, but above, the transition temperature, where the grain size is small, have higher levels of compressive strain.

Films deposited below the transition temperature and subsequently doped by implantation can be annealed at 850°C and yield

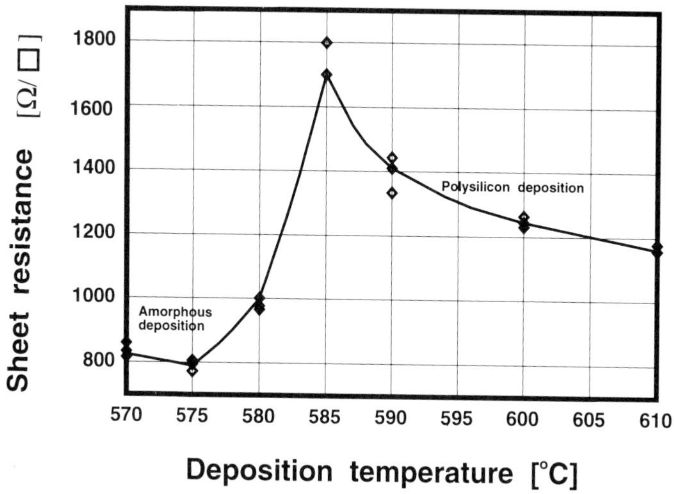

Figure 2.21. *Effect of deposition temperature on polysilicon sheet resistance for an implant dose of* $1 \cdot 10^{15} cm^{-2}$ *phosphorus, and a post-implant anneal of* $850^{\circ}C$ *for 30 min in* N_2.

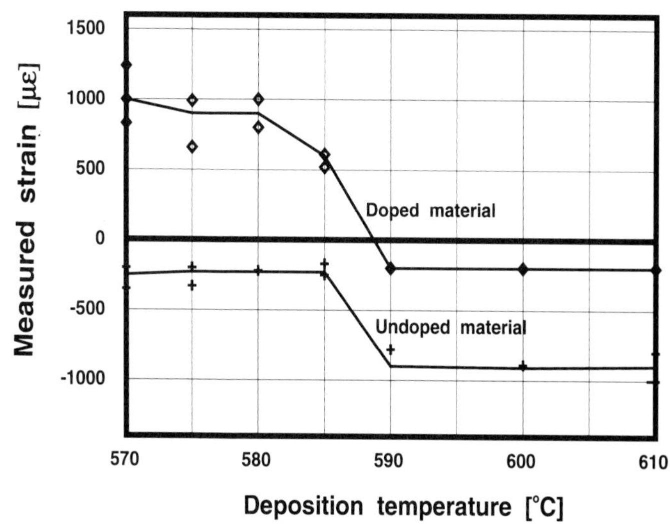

Figure 2.22. *Measured strain as a function of deposition temperature for the films described in Fig. 2.21, using oxidized polysilicon as a sacrificial layer.*

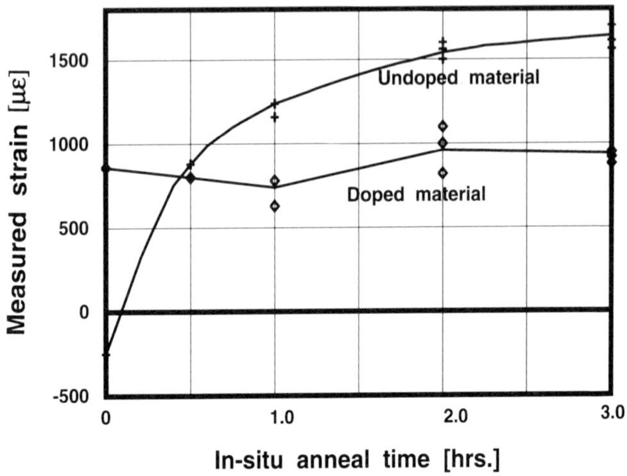

Figure 2.23. *Measured strain as a function of in-situ anneal time at a temperature of 600°C for films deposited at 575°C. Doped films were implanted with phosphorus at* $1 \cdot 10^{15} cm^{-2}$*, and annealed of 850°C for 30 min in* N_2.

satisfactory sheet resistance and tensile strain, whereas those deposited above this temperature remain in compressive stress even after doping and annealing. The ability to use such a low anneal temperature is highly desirable when integrating with electronic components.

As shown above, undoped and unannealed films are in compressive strain for all deposition temperatures. A low-temperature, 600°C, *in-situ* anneal for as little as 30 minutes is sufficient to have a significant effect on the stress levels in the films resulting in a change from a compressive strain of $\sim -300\mu\epsilon$ to a tensile strain of $\sim +1000\mu\epsilon$ for undoped films. Doped films are less influenced by the use of the *in-situ* anneal. The measured strain for these films is shown in Fig. 2.23. Once again, if the grain boundary is not formed during deposition, low temperatures can be used to remove any compressive strain. If a polysilicon structure is formed during deposition, such a low temperature anneal cannot be used as temperatures above 1000°C are required significantly change the struc-

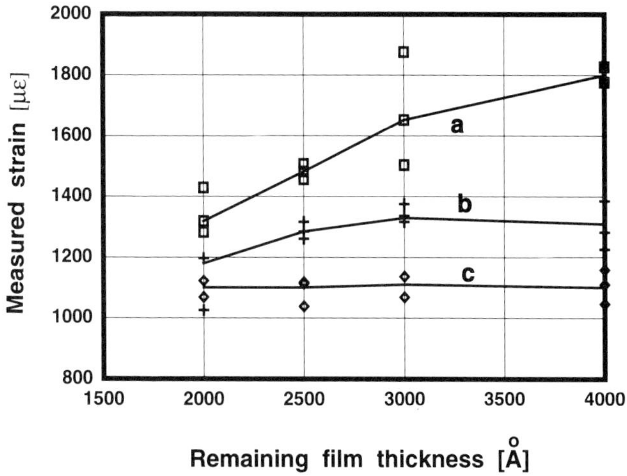

Figure 2.24. *Measured strain for thinned polysilicon layers for; (a) undoped films with no further processing, (b) undoped but with an 850°C anneal and (c) doped with phosphorus ($1{\cdot}10^{15}\,cm^{-2}$) and annealed at 850°C for 30 min.*

ture. It is therefore the formation of the grain boundary and not the grain size which is the main contributing factor in determining the stress levels. Transmission electron microscope (TEM) studies on these films showed that films deposited below the transition, and subsequently annealed, had similar grain sizes to those deposited near, but above the transition temperature, and also there was no significant difference in grain orientations. However there was a considerable difference in mechanical stress, being $+1000\mu\epsilon$ and $-500\mu\epsilon$. Further studies, including high-temperature annealing, revealed no correlation between grain size and mechanical strain.

A further consideration is the stress profile through the film. If there is a significant difference in mechanical stress through the film this may result in the bending up, or down, of single ended beams. Figure 2.24 shows the strain for 4000 Å films etched back to give an indication of the strain profile. This shows how the anneal has considerably reduced the profile and in the case of doped samples, no profile could be seen.

Table 2.5 Flexibility in the polysilicon process

Fixed parameters	Flexible parameter
Deposition pressure /temperature/gas flow (150 mTorr/570°C/45 sccm)	Film thickness (1000 Å–2 μm) Doping level ($5 \cdot 10^{14}$cm^{-2}–$5 \cdot 10^{15}$ cm^{-2}) Doping type (phosphorus – boron)
In-situ anneal (600°C, 60 min)	Implant energy (25 keV–120 keV) Anneal temperature/time (850°C–1100°C) (30–60 min)

The above experiments show how through optimization of two processing parameters, deposition temperature and *in-situ* anneal, a process can be developed with the desired mechanical properties while maintaining a low thermal budget. Although, as shown above, the *in-situ* anneal is not essential for the doped films, its incorporation ensures suitable mechanical properties for both doped and undoped films with a single standard process. If this process is to be combined with other structures, where the same polysilicon layer may also be used for other devices, with other requirements, it is important to establish the flexibility of the process. Flexibility in this context means processing parameters which may be varied, although two parameters are fixed, as mentioned above. The fixed and flexible parameters are shown in Table 2.5. The effects of these flexible parameters are discussed below. The last parameter, the anneal temperature, has been included to establish whether the polysilicon deposition can be performed at an earlier stage in the bipolar processing and still maintain the desired mechanical properties.

The first of the flexible processes is that of film thickness. The mechanical strain as a function of film thickness is shown in Fig. 2.25. The implanted energy for each film has been chosen to yield a range approximately in the middle of the film. This graph further shows a greater effect for the undoped polysilicon than for doped, which is similar to the strain profile measurements shown in Fig. 2.24.

Doping levels quoted in Table 2.5 are $5 \cdot 10^{14}\text{cm}^{-2}$–$5 \cdot 10^{15}\text{cm}^{-2}$. Lower doping levels yield high sheet resistance and are therefore

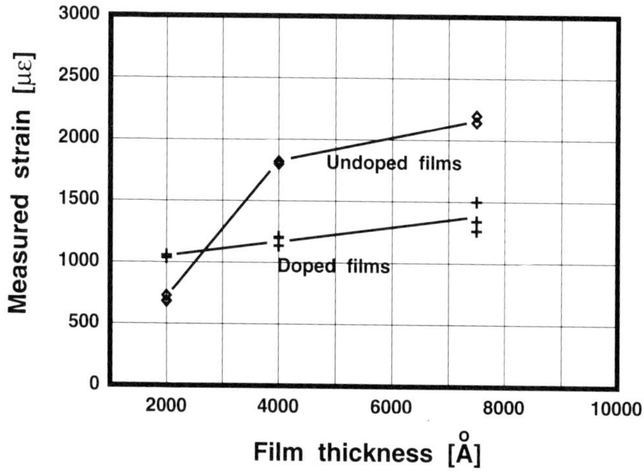

Figure 2.25. *Measured strain as a function of film thickness for undoped films and films implanted with phosphorus, $1{\cdot}10^{15}\,cm^{-2}$, and annealed at 850°C.*

not desirable for many applications. The sheet resistance of the polysilicon as a function of both phosphorus and boron doping levels is shown in Fig. 2.26. The mechanical strain is not greatly affected by doping level, as shown in Fig. 2.27, and this is an important degree of freedom for sensor applications. The strain as a function of implant energy is shown in Fig. 2.28. It shows a further degree of freedom.

The last parameter is that of the anneal temperature, which has two purposes in surface micromachining, that of removing the stress and activating the dopant. In the process described above, no stress anneal is required and therefore the anneal is only used to activate the dopant and temperatures of 850°C are sufficient for this purpose. If the micromachined device is to be combined with other devices, a further degree of freedom can be achieved in that the polysilicon processing can be placed at different stages of the process flow while maintaining satisfactory mechanical properties. For films deposited in a semi-amorphous form, strain levels reduce with increasing anneal temperature, crossing over

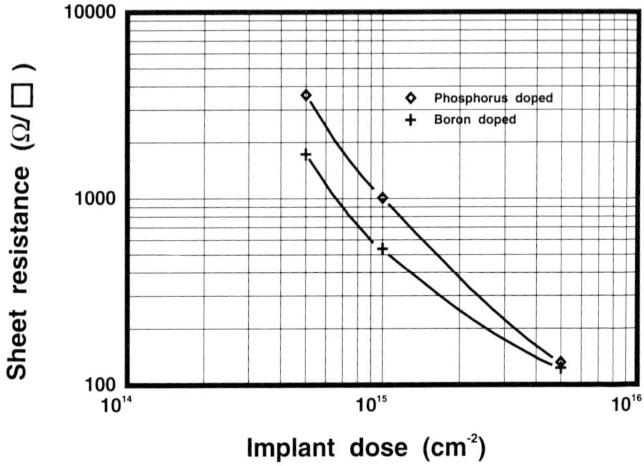

Figure 2.26. *Polysilicon sheet resistance as a function of implant dose for both phosphorus and boron implantation. Film thickness was 4000 Å and the post-implant anneal, 850°C for 30 min, in* N_2.

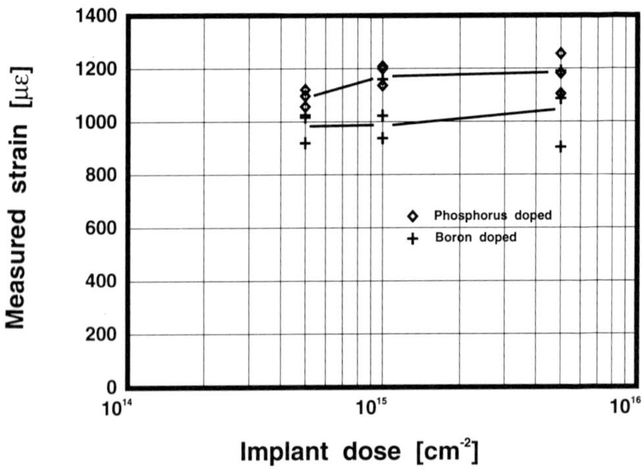

Figure 2.27. *Measured strain as a function of boron and phosphorus implant dose for a post-implant anneal of 850°C for 30 min in* N_2.

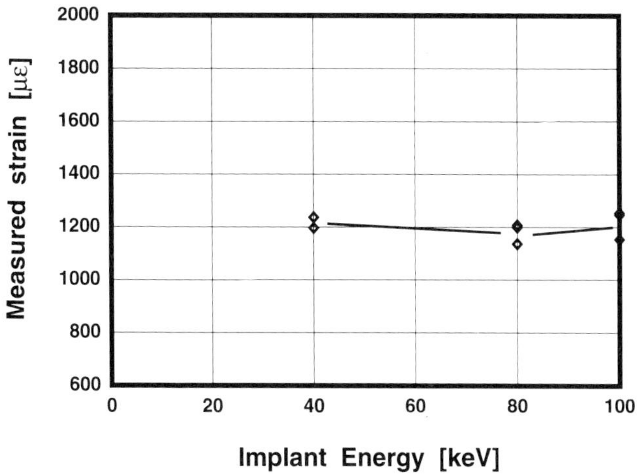

Figure 2.28. *Effect of implant energy on polysilicon strain for an implant dose of* $1 \cdot 10^{15} cm^{-2}$ *phosphorus and a post-implant anneal of 850°C for 30 min in* N_2.

to a compressive strain at anneal temperatures above 1000°C. For depositions above the transition temperature, the films are in compressive stress for all anneal temperatures, although the strain approaches zero as the anneal temperature is increased. Figure 2.29 shows the measured strains for films deposited at 570°C (below the transition) and 610°C (above the transition) as a function of anneal temperature after doping by phosphorus implantation at a dose of $1 \cdot 10^{15}\text{cm}^{-2}$, using both thermal oxide and PSG as a sacrificial material. This figure shows that if the deposition parameters are not optimized, high-temperature annealing is indeed required to remove the compressive stress.

It can be seen that both thermal oxide and PSG are suitable for use as a sacrificial layer, although PSG has the great advantage of a high etch rate. A further consideration is the effect of changes in the PSG processing on the polysilicon characteristics. The first consideration is the influence on sheet resistance and this is shown in Fig. 2.30. This shows that, due to out-diffusion during annealing, higher phosphorus levels in the PSG resulted in

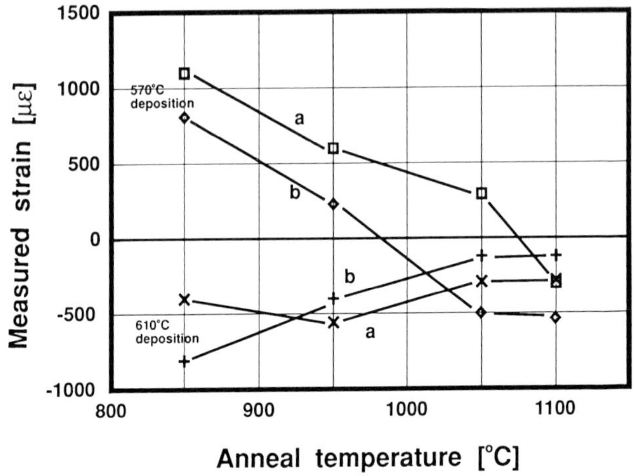

Figure 2.29. *Effect of post-implant anneal temperature on polysilicon strain for an implant dose of* $1 \cdot 10^{15} cm^{-2}$, *phosphorus for films deposited on both oxidized polysilicon, curves a, and PSG, curves b.*

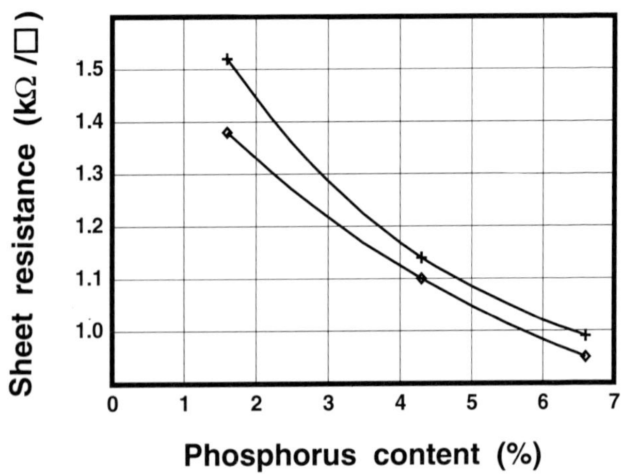

Figure 2.30. *Effect of phosphorus out-diffusion from the PSG on polysilicon sheet resistance.*

Table 2.6 Factors affecting mechanical stress in nitride films

LPCVD nitride	PECVD nitride
Gas flow ratio	Gas flow ratio
Deposition temperature	Deposition temperature
Pressure	Pressure
	Plasma
	Power
	Frequency
	Duty-cycle

a lower polysilicon sheet resistance. Generally, a low polysilicon sheet resistivity is required and such an out-diffusion of phosphorus from the PSG would not give rise to any problems. However, for some applications, PSG with high phosphorus contents should be avoided as this out-diffusion might lead to leakage currents between polysilicon structures and other layers, due to doping of insulating layers. The stress levels are relatively unaffected.

This process shows flexibility in dopant type, implantation dose and energy. As expected the post-implant anneal has less flexibility. An anneal of 850°C, 30 min in N_2 yields suitable mechanical properties but increasing the temperature decreases the tensile strain and eventually leads to compressive strain. Therefore the best use of this polysilicon is to deposit at the end of the bipolar process, but before interconnect processing.

2.5.3 The micromechanical layers II — nitride

The second main material used as a micromechanical layer is silicon nitride, which is a mechanically strong material with low optical absorption. The stress levels in the material, as with polysilicon, are highly dependent upon both deposition parameters and subsequent thermal processing. Deposition can be performed by either LPCVD or PECVD.

Control of mechanical stress is quite different for the two systems and the important factors are listed in Table 2.6.

Plasma enhanced systems operate at temperatures compatible

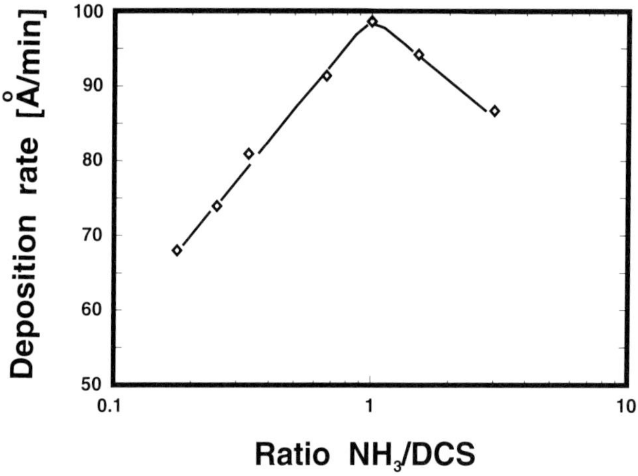

Figure 2.31. *Deposition rate of silicon nitride as a function of ammonia/DCS ratio.*

with aluminum and therefore it is possible to use PECVD for fabrication of micromechanical structures after the aluminum has been deposited. If, however, this is not required, LPCVD is a suitable option. This description will concentrate on LPCVD deposited films.

The films are usually deposited using dichlorosilane (DCS) and ammonia (NH_3), usually at a temperature range of 750°C to 850°C and pressures of 80 mTorr to 300 mTorr. Within the deposition parameter range described above, the mechanical strain can vary from $100\,\mu\epsilon$ to $3600\,\mu\epsilon$, tensile. Higher deposition temperatures have been used to yield lower stress [2.55] but this would create an addition to the thermal budget and is therefore undesirable. Therefore, the 850°C deposition represents the best option. Therefore, the gas flows and pressure have to be optimized in terms of mechanical stress. In this section the optimization of each is described in turn.

The deposition rate and refractive indices are shown in Figs 2.31 and 2.32, respectively. The deposition rate is greatly dependent upon gas flow reaching a peak for equal flows. Refractive index

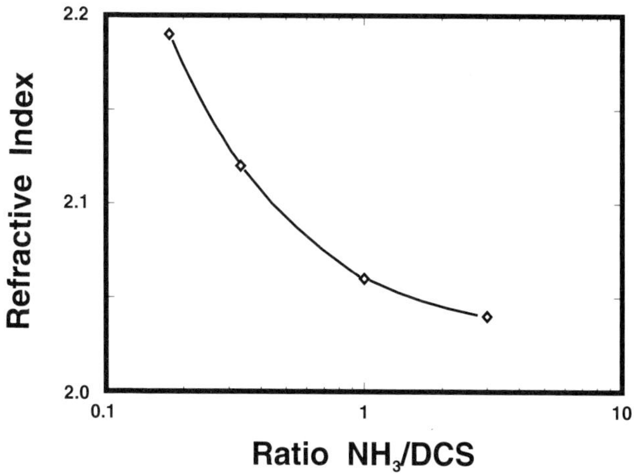

Figure 2.32. *Refractive index of silicon nitride as a function of ammonia/DCS ratio.*

Table 2.7 Etch rate of SiN in HF

DCS	NH_3	Etch rate BHF ($Å min^{-1}$)	20% HF ($Å min^{-1}$)	40% HF ($Å min^{-1}$)
170	30	3.1	17	30
150	50	3.6	19	30
100	100	4.5	26	47
50	150	5.5	32	52

shows higher values for higher DCS flow because of the increased silicon content. The refractive index is not of mechanical importance but when optics and mechanics are combined, such as for interferometers [2.56], this is an important consideration.

A further consideration for nitride, which was not a problem for polysilicon, is the etch rate in HF. The etch rates in BHF, 20%HF and 40%HF are given in Table 2.7. It can be seen that the etch rate reduces as the silicon content is increased. Higher silicon content leads to lower strain levels, as shown in Fig. 2.33. These films were deposited at a temperature of 850°C. The high

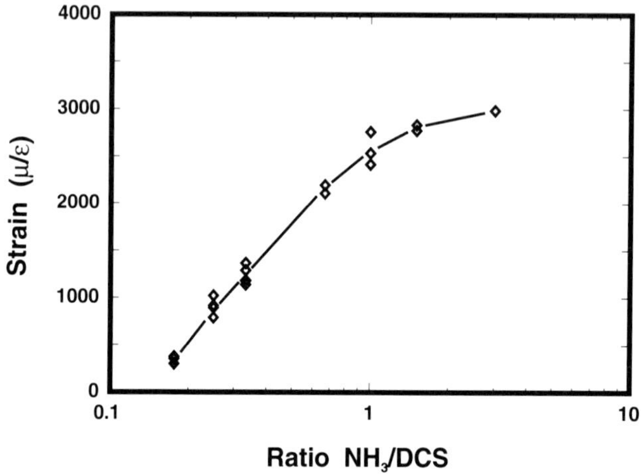

Figure 2.33. *Silicon nitride strain as a function of ammonia/DCS ratio.*

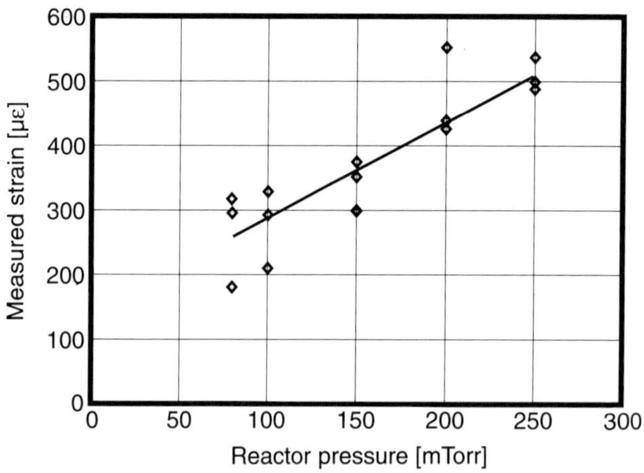

Figure 2.34. *Silicon nitride strain as a function of deposition pressure.*

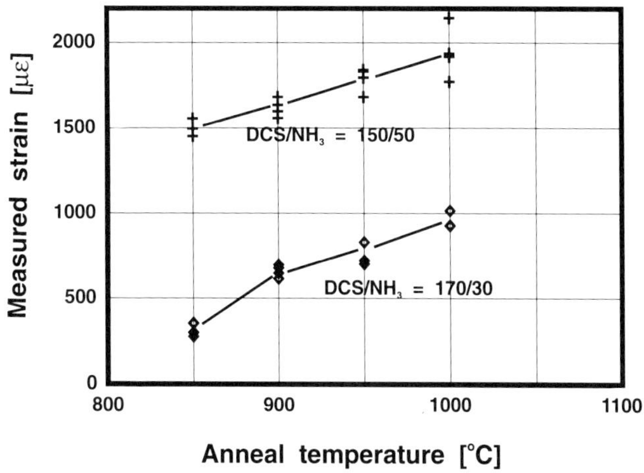

Figure 2.35. *Silicon nitride strain as a function of post-deposition anneal temperature, for films deposited at a pressure of 150 mTorr and DCS/NH_3 ratios of (a) 170/30 and (b) 150/50.*

silicon content shows mechanical advantages but can present two problems: (1) the variations in the film quality and (2) DCS yields greater problems for keeping the system clean. The strain levels can be further reduced by increasing the deposition temperature [2.55], but an increase in temperature would not be desirable as it would increase the thermal budget. A reduction in pressure yields only a small reduction in strain as shown in Fig. 2.34.

All the above strain measurements are for nitride films which have had no further thermal treatment above the deposition temperature. Further annealing can have a considerable effect on the mechanical properties of the films, as shown in Fig. 2.35 [2.57]. This considerable effect of anneal temperature is a possible limitation for this material.

2.5.4 Planarization/step coverage

Each time a layer is deposited on the surface of the wafer this creates a step which, if several layers are deposited, can create two problems: (1) step coverage of upper layers and (2) uniformity of resist.

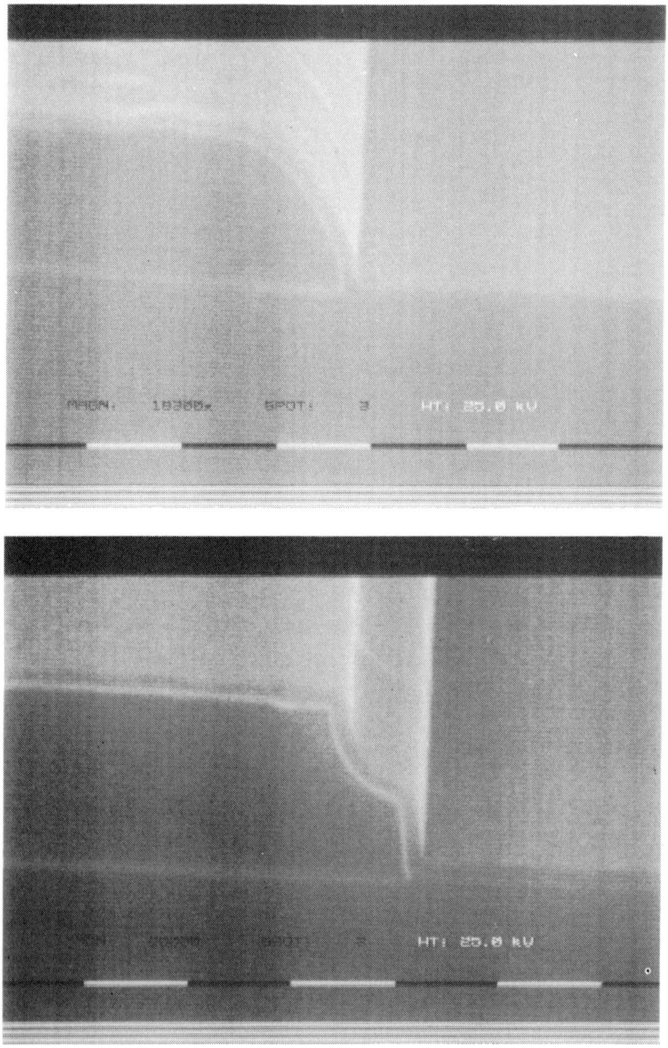

Figure 2.36. *SEM photographs of step modification using combinations of plasma and wet etching.*

Step coverage for polysilicon or nitride does not usually create problems but, if necessary, the step shape can be modified as shown in Fig. 2.36 [2.58]. In all cases the step coverage for polysilicon is sufficiently good. However, aluminum is known to have poor step coverage. This can be overcome by using a staircase structure where the aluminum climbs one layer at a time. The second problem is that of resist spinning. Once large steps have been formed on the wafer, subsequent masking steps will have to use a thicker resist to ensure sufficient cover on the outer part of the wafer. This can ensure reasonable resist cover but limits the feature size that can be defined, and is therefore not suitable for defining small contact windows.

One possible solution to the problems described above is to planarize the wafer to fill cavity area with oxide. Both PSG and BPSG have been used for such purposes. The advantage of BPSG is its relatively low reflow temperature [2.59]; however the etch rate is considerably lower than that of PSG. An alternative technique is to use a plasma etch where the etch rate of PSG is the same as the resist [2.60], since the resist is spun on and yields a planar surface. This does however require thicker PSG layers and accurate back-etching.

Rather than etching back, selective epitaxial growth (SEG) provides the opportunity of growing the epitaxial layer around the PSG [2.61]. Provided the PSG areas are aligned in the [100] direction the epitaxial growth yields a planar surface on which the mechanical layers can be deposited. The same epitaxial layer can also be used for the electronics [2.62].

2.5.5 Contact holes

In the standard electronic processing, the electronics and the interconnect layer are separated by a thermal silicon dioxide layer. When etching the sacrificial layer for patterning and freeing the structures in hydrofluoric acid (HF), this silicon dioxide isolation layer is also removed. A further problem is that the micromechanical layers which are deposited on top of the oxide must be patterned. The etch must therefore stop on the oxide, and therefore each over-etch thins the isolation. If a thin nitride layer is deposited on this oxide it serves as an etch-stop. It must therefore be thick enough to survive all etch steps.

The contact holes that have to be opened in the nitride layer and the underlying oxide layer provide access to this oxide layer

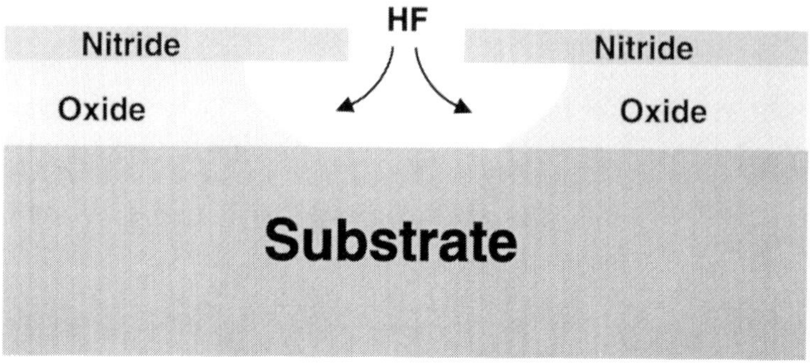

Figure 2.37. *Under-etching of contact windows opened before sacrificial etching.*

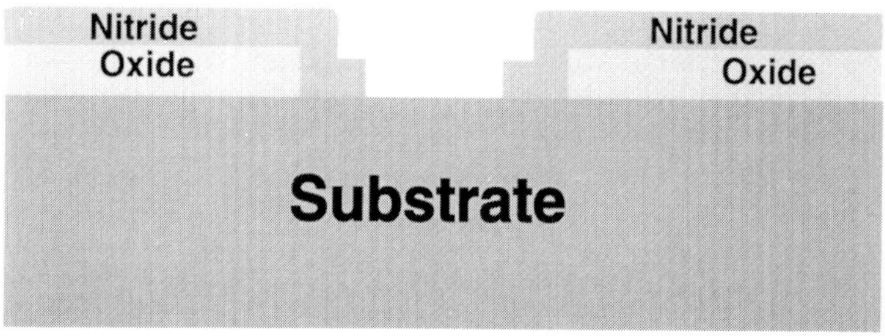

Figure 2.38. *Protection of opened contact windows using double contact window etching.*

(Fig. 2.37). If no measures are taken, the nitride layer will be strongly under-etched during the sacrificial etch, which may cause problems. One solution is to use a double contact window where the oxide is completely protected by nitride as shown in Fig. 2.38.

A further problem with contact holes is the forming of small windows when large steps have been formed on the chip. Large steps result in a non-uniformity of resist thickness and thus the contact windows cannot be accurately defined.

2.5.6 Aluminum interconnect

Further problems are caused by the aluminum interconnect layer. The etching of the sacrificial oxide layer is performed in hydrofluoric acid (HF) which also attacks the metal. Therefore, it is necessary to protect the aluminum layer during this etch or to sputter the aluminum *after* the structures have been freed. Usually it is not possible to spin-coat the free standing structures with photoresist, as this would result in high risk of breakage. However, the sputtered and unpatterned aluminum layer supports the structures sufficiently to make the spinning of resist and patterning of this layer possible after the structures are released. The second problem is caused by the 2% silicon that is added to the aluminum, to avoid spiking of the emitter junctions and reduce electromigration. After patterning the aluminum, small grains of silicon remain on the wafer. In a normal electronic process, these grains are removed using a polysilicon dip-etch. If the top micromachined layer is polysilicon, such a dip-etch cannot be used. If the polysilicon layers are not released before the aluminum deposition, the polysilicon can be protected using oxide or nitride. If possible an oxide should be used, as this will automatically be removed during sacrificial etching. If the structures are already released, the structure can be oxidized before deposition of the interconnect. This will protect the polysilicon layers, but some thickness will be lost, it can cause extra sticking problems and it will considerably add to the thermal budget.

Considering the above argument, it is preferable to form the aluminum before sacrificial etching and to protect any polysilicon, from the dip-etch, with a deposited oxide layer. The best protection for the electronics during etching must be established; this is discussed below.

2.5.7 Sacrificial etching

Sacrificial etching is the step where the sacrificial layer is removed from under the micromechanical layers to produce free standing structures. It is important to ensure that the etch selectivity is high and the electronics is protected. Using high HF concentrations yields a high under-etch rate and therefore large structures can easily be etched. If, as mentioned above, the electronics is protected by resist, BHF is the most suitable etchant. An alternative solution is to use a plasma-deposited nitride layer to protect the electronics. This would have to be sufficiently thick to withstand the HF etching. The remaining layer would have to removed using plasma etching after sacrificial etching. This may, however, present significant problems with over-etching of mechanical layers during the removal of the nitride.

The sacrificial etching mechanism is also an important consideration and there have been several recent studies on this topic [2.63–2.65]. The etch rate can be enhanced adding HCl to the HF solution [2.66].

An additional problem is that of sticking. When structures are released they remain free while remaining in the etchant. If the structures are cleaned in water and allowed to dry, capillary forces will pull the structures down to the substrate. This can create severe problems for fabricating surface micromachining devices and there have been several studies into the effect [2.67–2.68]. Therefore considerable care must be taken in the drying process. This process is further complicated when resist is present. There are two main techniques. The first is to immerse the sample in alcohol and dry the structure on a hot plate. The alcohol, with its low surface tension, is able fill the cavity and rapidly evaporates at the elevated temperature. An alternative technique is known as freeze dry. The technique makes use of sublimation to release the structures. This can be achieved by lowering the pressure of a sealed chamber containing the wafer [2.48, 2.69], but this technique, although effective, has the disadvantage of being complicated. A simpler freeze dry technique involves the use of cyclohexane, which has a freeze temperature of +7°C [2.70]. After etching the samples are rinsed in water (with water or alcohol to maintain a wet surface), and then the resist is removed in acetone. The sample is immersed in cyclohexane and then reduced to a temperature of 5°C in a nitrogen flow. The cyclohexane sublimates and the sample is gradually brought back to room temperature.

2.5.8 Compatible surface micromachining process

In this section the potential problems in the combining of electronics with surface micromachining structures are continued. To ensure the desired mechanical properties of the devices, these depositions should be performed after the high-temperature steps of the electronics processing. By maintaining an anneal temperature of 850°C the surface micromachining processing does not adversely affect the electronics [2.6,2.52]. A further important consideration is the protection of the electronics during the etching of the mechanical layers. This is usually performed by plasma etching and therefore selectivity over the underlying layers is extremely important. Such selectivity can be achieved by careful consideration of the plasma gases [2.71].

There are two remaining problems — these are the sacrificial etching and the planarity of the wafer surface.

The sacrificial etching can be solved by either protecting the electronics with nitride or by using resist [2.34,2.72]. Both layers are attacked by HF. In the case of resist, the etching should be performed in BHF which has the disadvantage of increasing the etching time. However, resist has the advantage that it can easily be completely removed after etching. Using nitride permits the use of higher HF concentrations, although the layer should be thick enough to survive the etching. The disadvantage of this technique is the layer has to be removed after etching to make contact to the aluminum. This can be achieved using plasma etching.

Once a compatible surface micromachining process has been developed many new opportunities for sensor structures arise. Structures were extremely small capacitance changes must be measured [2.73] and where mechanical devices are used as electrical components [2.74]. The compatible surface micromachining process, developed at Delft University, is illustrated in Fig. 2.39 [2.34]. Firstly the bipolar devices are fabricated up, but not including, the contact windows. A nitride layer is deposited over the wafer. This layer serves to protect the electronics during subsequent etching of the micromachined layers (Fig. 2.39a). The thickness must therefore be sufficient to withstand any over-etching. The sacrificial layer is then deposited and patterned followed by the formation of contact windows to the substrate, if necessary, as shown in Fig. 2.39b. The micromechanical layer is then deposited and patterned, as shown in Fig. 2.39c. If the upper micromechanical layer is polysilicon, an oxide layer is deposited. Using PECVD ensures no additional thermal budget from this step. This layer will serve as a protective layer at a later stage. Contact windows and aluminum deposition are performed using standard processing, as shown in Fig. 2.39d.

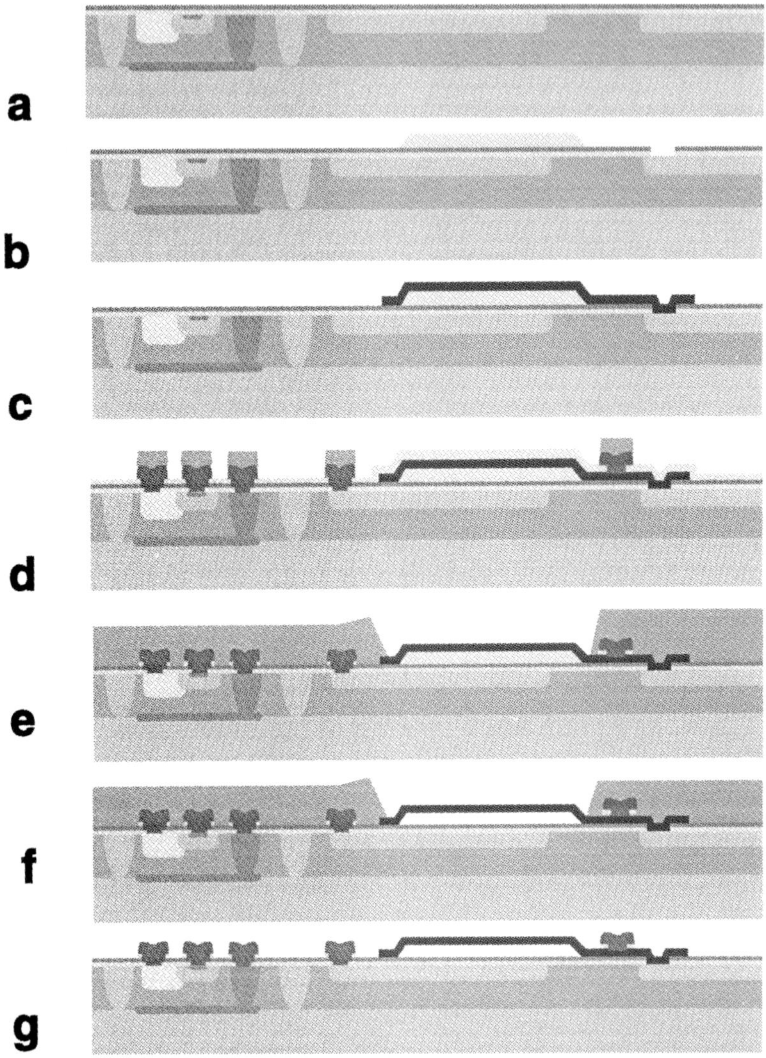

Figure 2.39. *The compatible surface micromachining process developed at Delft University [2.34]. (a) Up to contact window etching, (b) deposition and patterning of sacrificial layer followed by contact window for micromechanical layer, (c) deposition and patterning of micromechanical layer, (d) deposition of PECVD oxide, followed by aluminum sputtering and patterning, (e) deposition of resist protective layer, (f) sacrificial etching and (g) removal of resist.*

After aluminum etching (which contains 2% Si) a polysilicon dip-etch is performed to remove the silicon grains left on the surface. After the polysilicon dip-etch the PECVD oxide is removed, before removal of the aluminum mask. Sacrificial etching is performed in HF and therefore the aluminum must be protected. This protection is achieved using resist (Figs 2.39e and 2.39f) which, when BHF is used, can is sufficient to protect the underlying layers. After sacrificial etching the resist is removed in acetone and the wafer is dried using the process described in section 2.5.7, resulting in the final structure shown in Fig. 2.39g.

2.6 LIGA process

The LIGA process is a quite different approach to surface micromachining to achieve high aspect ratios. Basically the process combines X-ray lithography with thick resist layers and electroplating metal layers [2.75] It can also be used with sacrificial layers which is know as sacrificial LIGA [2.76]. Of particular interest for micromachining is the use of a sacrificial layer to fabricate free standing structures. An outline of this process is shown in Fig. 2.40.

This process cannot be considered to be compatible with a clean line. However the process requires no high temperatures and can be performed at the end of the processing for the electronics, outside the cleanroom.

2.7 Wafer-to-wafer bonding

A further option used to create sealed cavities with bulk micromachined devices is to bond an additional chip, either silicon or glass, to the sensor chip. Bonding is also frequently used to obtain silicon on oxide devices [2.77–2.79]. A common example of glass to silicon bonding is for the bulk micromachined accelerometer. In this case the glass plate serves several functions:

1. To provide a seal and the desired damping
2. In the case of capacitive readout, a metal plate on the glass serves as one plate of the capacitor.
3. As overload protection.

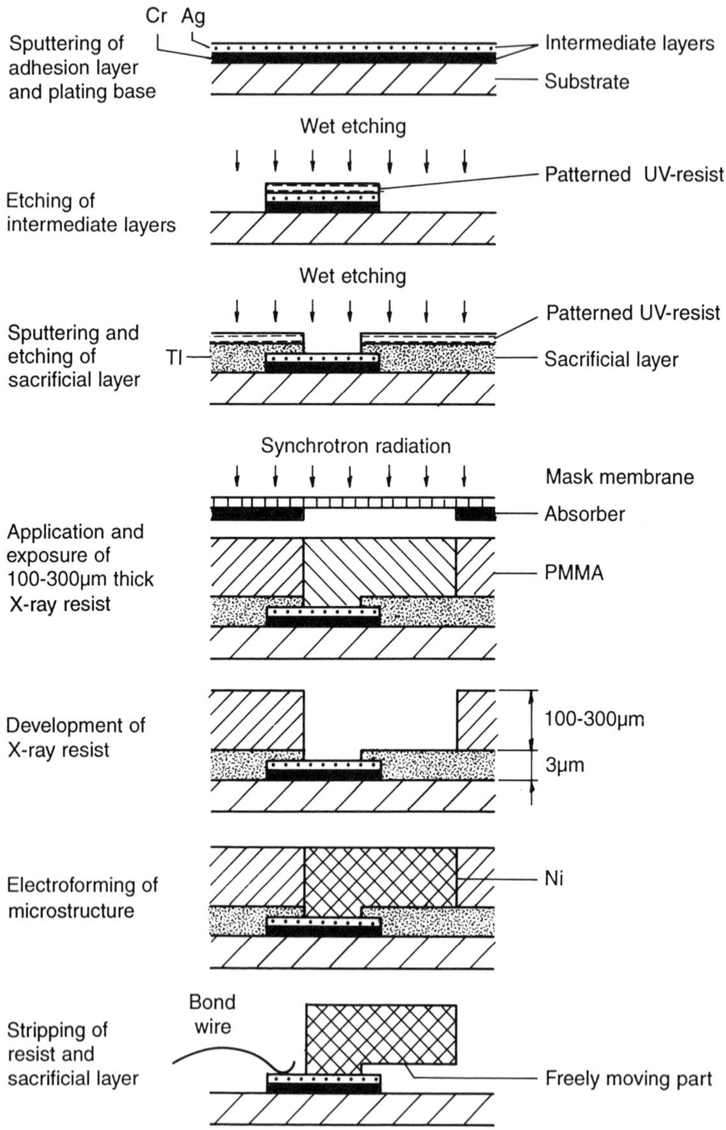

Figure 2.40. *LIGA process sequence taken from Becker et al. [2.75].*

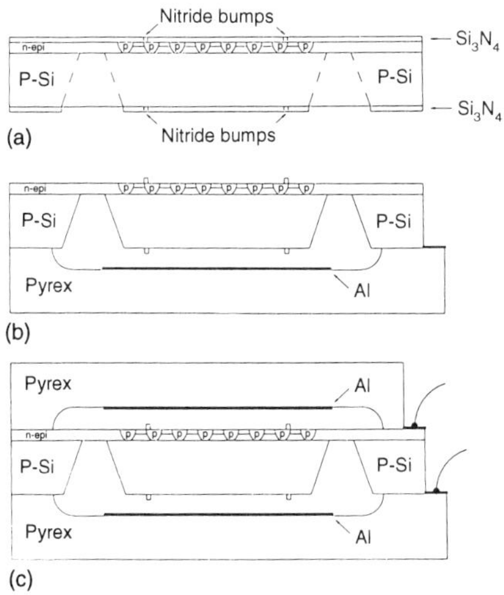

Figure 2.41. *Fabrication of a triple glass/silicon/glass structure.*

The fabrication of this structure is shown in Fig. 2.41 [2.13].

The bonding is achieved by anodic bonding, which requires an elevated temperature <450°C and high voltage 1000 V. This technique is commonly used with fully processed wafers. The temperature is low enough to present no problems for the aluminum and the voltage is applied between the glass and the wafer and therefore there is no problem heating of the aluminum through high currents.

Silicon direct wafer bonding is a technique for fusing wafers by heat and no additional layer is required to aid adhesion. Both wafers should be polished and hydrophilic (thus containing a high density of OH groups on the surface — this can be achieved by immersing in fuming nitric acid). The wafers are brought together at room temperature by either electrostatic of mechanical means. The wafers are then annealed to increase the strength of the bond. Early techniques used high temperatures and high pressures to bond the wafer [2.79]. Although temperatures of above 1000°C are needed, the short time required does not result in excessive out-diffusion of dopants. However, the mechanical properties may be changed and wafers with aluminum cannot be used. Lower temperatures can be used but this results in a lower bond strength, as shown in Fig. 2.42 [2.1].

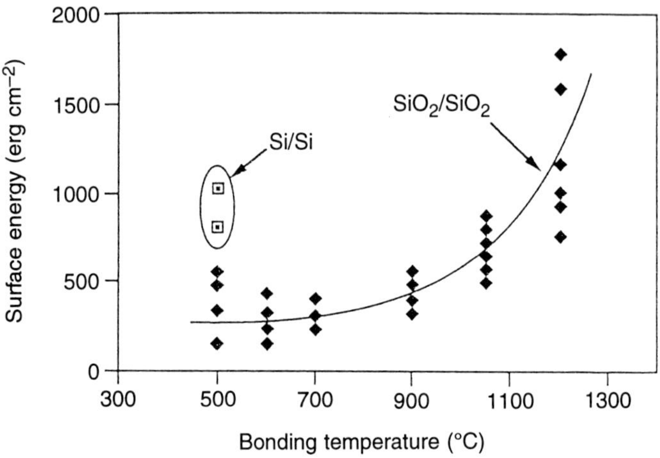

Figure 2.42. *Bond strength as a function of bond temperature. Taken from Ristic [2.1].*

Low temperature wafer-to-wafer bonding has been achieved using an intermediary layer such as polymer glue, low-temperature glass [2.80], or gold [2.81]. Bonding with a glue is a simple technique, but has several drawbacks. Reproducibility and contamination due to outgassing are the main disadvantages. However, this technique has been successfully applied to the fabrication of SOI devices [2.82]. Boron-doped glass has been applied to low-temperature bonding. This softens at relatively low temperatures [2.83]. Bonding takes place at 400°C and 20–200 V and takes about 10 min. The problem with boron-doped glass is the sensitivity to phosphorus contamination which considerably increases the bonding temperature. A recent alternative is a sodium silicate glass which bonds at 200°C for 2 hours [2.84] yielding high bond strength. Alternatively, materials such as gold have been used for a through-wafer bonding process [2.81]. This process uses a 300 Å titanium/1200 Å gold layer combination deposited on an oxidized wafer. The titanium is used to improve the adhesion. Bonding was performed at temperatures between 350°C and 400°C with a 100 g distributed weight. This low-temperature process is thus, in terms of thermal budget, compatible with standard processing, although the use of gold means that, to avoid contamination, the bonding must be performed at the end of the process.

There is a range of structures in the literature which uses silicon-to-silicon bonding and some examples are given in [2.85–2.87].

2.8 Conclusions

This chapter has considered many of the problem areas in the integration of electronics and micromachined devices. Careful consideration is therefore necessary on the effects of each process requirements on the other devices. Furthermore, compatibility with the processing environment must also be considered.

In the case of surface micromachining, the thermal budget is an important consideration. A process has been described here has a total thermal budget compatible with standard electronics processing. The additional problems of step coverage and protecting the electronics during sacrificial etching have been addressed, resulting in a fully compatible process.

Bulk micromachining usually presents no thermal budget problems since low temperatures are used. As with surface micromachining, the electronics must be protected during etching. In many applications the final etch must be performed using plasma etching from the front-side. Therefore extra care must be taken in the handling of the wafers since the devices are mechanically weakened by the back-side etching.

In many applications some form of wafer or glass bonding is used. This often requires high temperatures or high voltages. The high temperatures can present problems for the electronics. The high voltages can be accommodated, with care.

With careful consideration of the thermal budget and compatibility of etchants and deposited layers, a wide range of smart sensors and actuators with mechanical components and full electronic circuitry (both bipolar and CMOS) can be fabricated.

2.9 References

[2.1] L.J. Ristic ed., *Sensor technology and devices*, Artec House, London, UK.

[2.2] L. J. Ristic, CMOS technology: a base for micromachining, *Microelectronics J.*, 20 (1989) 153.

[2.3] H. Baltes, CMOS as a sensor technology, *Sensors and Actuators*, A37–38 (1993) 51–56.

[2.4] L.K. Nanver, E.J.G. Goudena and H. W. van Zeijl, DIMES-01 baseline BIFET process for smart sensor experimentation, *Sensors and Actuators*, A36 (1993), 139–147.

[2.5] P.M. Sarro, Sensor technology strategy in silicon, *Sensors and Actuators*, A31 (1992) 138–143.

[2.6] K.M. Mahmoud and R.F. Wolffenbuttel, Compatibility between bipolar read-out electronics and microstructures in silicon, *Sensors and Actuators*, A31 (1992) 188–199.

[2.7] K.E. Petersen, Silicon as a mechanical material, *Proc. IEEE*, 70 (1982) 420–457.

[2.8] K.E. Bean, Anisotropic etching of silicon, *IEEE Trans. Electron Devices*, ED-25 (1978) 1185–1193.

[2.9] A.W. van Herwaarden and P.M. Sarro, Floating membrane thermal vacuum sensor, *Sensors and Actuators*, 14 (1988) 259–268.

[2.10] P.M. Sarro, A.W. van Herwaarden and W. van der Vlist, A silicon-silicon nitride membrane fabrication process for smart thermal sensors, *Sensors and Actuators* A41–42 (1994) 666–671.

[2.11] S.K. Clark and K.D. Wise, Pressure sensitivity in anisotropically etched thin diaphragm pressure sensors, *IEEE Trans. Electron Dev.*, ED-26 (1979) 1857–1896.

[2.12] H.V. Allen, S.C. Terry and D.W. de Bruin, Accelerometer systems with self-testable features, *Sensors and Actuators*, 20 (1989) 153–161.

[2.13] R.P. van Kampen, M.J. Vellekoop, P.M. Sarro and R.F. Wolffenbuttel, Application of electrostatic feedback to critical damping of an integrated silicon capacitive accelerometer, *Sensors and Actuators*, A43 (1994) 100–106.

[2.14] V. Gass, B.H. van der Schoot, S. Jeanneret and N.F. de Rooij, Integrated flow regulated silicon micropump, *Sensors and Actuators*, A43 (1994) 335–338.

[2.15] H.T.G. van Lintel, F.C.M. van de Pol and S. Bouwstra, A piezoelectric micropump based on micromachining of silicon, *Sensors and Actuators*, 15 (1988) 153–167.

[2.16] T.A. Kwa and R.F. Wolffenbuttel, Integrated grating/detector array fabricated in silicon using micromachined techniques, *Sensors and Actuators*, A31 (1992) 256–266.

[2.17] K.M. Mahmoud, R.P. van Kampen, M.J. Rutka and R.F. Wolffenbuttel, A silicon integrated smart pressure sensor, *Proceedings Transducers'93*, Yokohama, Japan, pp. 217–220.

[2.18] C. Hakan Özdemir and J.G. Smith, New phenomena observed in electrochemical micromachining of silicon, *Sensors and Actuators*, A31 (1992) 87–93.

[2.19] S. Tatić-Luić and Y.-C. Tai, Novel extra accurate method for two-sided alignment on silicon wafers, *Sensors and Actuators*, A41–42 (1994) 573–577.

[2.20] T.A. Kwa and R.F. Wolffenbuttel, Potential defined etch mask for silicon micromachining, *Proceedings Eurosensors VII*, San Sebastian, Spain, p. 370.

[2.21] A.D. Khazan, *Transducers and their elements*, PTR Prentice Hall, 1994.

[2.22] J.B. Price, Anisotropic etching of silicon with KOH-H_2O-Isopropyl alcohol, in *Semiconductor Silicon* ed. H.R. Huff and R.R. Burgess, Princeton, NJ, Electrochemical Society Proceedings, 1973, p. 339.

[2.23] H. Seidel, L. Csepregi, A. Heuberger and H. Baumgärtel, Anisotropic etching of crystalline silicon in alkaline solution; Part I Orientation dependence and behavior of passivation layers; Part II Influence of dopants, *J. Electrochem. Soc.*, 137 (1990) pp 3612–3626.

[2.24] M.J. Vellekoop, C.C.G. Visser, P.M. Sarro and A. Venema, Compatibility of zinc-oxide with silicon IC processing, *Sensors and Actuators*, A21–A23 (1990) 1027–1030.

[2.25] B. Peurs and W. Sansen, Compensation structures for convex corner micromachining in silicon, *Sensors and Actuators*, A21–23 (1990) 1036–1041.

[2.26] R.P. van Kampen and R.F. Wolffenbuttel, Effects of <110> oriented corner compensation structure on membrane quality and convex corner integrity in (100) silicon using aqueous KOH, *Proceedings MME 94*, Pisa Italy.

[2.27] S. Boustra, R. Legtenberg, H.A.C. Tilmans and M. Elwenspoek, Resonating microbridge mass flow sensor, *Sensors and Actuators*, A21–23 (1990) 332–335.

[2.28] A.J. Jacobs-Cook and M.E.C. Bowen, Planar fibre-optic interface for silicon microresonators, *Sensors VI: Technology, Systems and Applications*, Manchester, U.K., 1993, pp 147–154.

[2.29] L. Rosengren, L. Smith and Y. Bäcklund, Micromachined optical planes and reflectors in silicon, *Sensors and Actuators*, A41–42 (1994) 330–333.

[2.30] J. Brugger, N. Blanc, Ph. Renaud and N.F. de Rooij, Microlever with combined integrated sensors/actuators function for scanning force microscopy, *Sensors and Actuators*, A43 (1994) 339–345.

[2.31] J. Foerster, Y.X. Li, M. Bartek, P.J. French and R.F. Wolffenbuttel, Fabrication of recessed micro-tips in silicon for sensor applications, *Proceedings MME 94*, Pisa, Italy, 1994.

[2.32] K.A. Shaw, Z.L. Zhang and N.C. MacDonald, SCREAM I: a single mask, single-crystal silicon, reactive ion etching process for microelectromechanical structures, *Sensors and Actuators*, A40 (1994) 63–70.

[2.33] Y.X. Li, P.J. French, H. Schellevis and R.F. Wolffenbuttel, SIMPLE-A novel fabrication technique for silicon micromachining by a single plasma etching step, *Proceedings Eurosensors 94*, Toulouse, 26–28 September 1994, p. 112.

[2.34] B.P. van Drieënhuizen, J.F.L. Goosen, P.J. French, Y.X. Li, D. Poenar and R.F. Wolffenbuttel, Surface micromachined module compatible with BiFET electronic processing, *Proceedings Eurosensors 94*, Toulouse, France, September 1994, p. 108.

[2.35] W. Lang, A. Richter, K. Maruszyk, G. Weimann and H. Sandmaier, Application of porous silicon as a sacrificial layer, *Proceedings Transducers'93*, Yokohama, Japan, pp. 202–205.

[2.36] M. Ataka, A. Omodaka and H. Fujita, A biomimetic micro motion system–a ciliary motion system, *Proceedings Transducers'93*, Yokohama, Japan, pp. 38–41.

[2.37] K. Shimaoka, O. Tabata, M. Kimura and S. Sugiyama, Micro-diaphragm pressure sensor using polysilicon sacrificial layer etch-stop technique, *Proceedings Transducers'93*, Yokohama, Japan, pp. 632–635.

[2.38] O. Tabata, R. Asahi, N. Fujitsuka, M. Kimura and S. Sugiyama, Electrostatic driven optical chopper using SOI wafer, *Proceedings Transducers'93*, Yokohama, Japan, pp. 124–127.

[2.39] B. Diem, M.T. Delaye, F. Michel, S. Renard, and G. Delapierre, SOI (SIMOX) as a substrate for surface micromachining of single crystalline silicon sensors and actuators, *Proceedings Transducers'93*, Yokohama, Japan, pp. 223–236.

[2.40] M.-A. Grétillat, C. Linder and N.F. de Rooij, Multilayer polysilicon resonators including shielding for excitation and detection, *Proceedings Transducers'93*, Yokohama, Japan, pp. 292–295.

[2.41] D. Poenar, P.J. French, R. Mallée, P.M. Sarro and R.F. Wolffenbuttel, PSG for surface micromachining, *Sensors and Actuators*, 41 (1994) 304–309.

[2.42] Y. Onuma and K. Sekiya, Piezoresistive properties of polycrystalline silicon thin films, *Jpn. J. Appl. Phys.*, 11 (1974) 420–421.

[2.43] J.Y.W. Seto, The electrical properties of polycrystalline silicon films, *J. Appl. Phys.*, 46 (1975) 5247–5254

[2.44] M.M. Mandurah, K.C. Saraswat, T.I. Kamins, A model for conduction in polysilicon: Part 1-Theory, Part 2 - Comparison of theory and experiment, *IEEE Trans. Electron Devices*, ED28 (1981) 1163–1176.

[2.45] H. Mikoshiba, Stress-sensitive properties of silicon gate MOS devices, *Solid State Electronics*, 24 (1981) 221–232.

[2.46] P.J. French and A.G.R. Evans, Piezoresistance in polysilicon and its applications to strain gauges, *Solid State Electronics*, 32 (1989) 1–10.

[2.47] S.P. Murarka and T.F. Retajczyk, Jr., Effect of phosphorus doping on stress in silicon and polycrystalline silicon, *J. Appl. Phys.*, 54 (1983) 2069–2072.

[2.48] H. Guckel, D.W. Burns, C.R. Rutigliano, D.K. Showers and J. Uglow, Fine grained polysilicon and its application to planar pressure transducers, *The Fourth International Conference on Solid-State Sensors and Actuators*, Tokyo, Japan, 1987, Digest of Technical Papers, p. 277.

[2.49] R.S. Hajab and R.S. Muller, Residual strain effects on large aspect ratio microdiaphragms, *Proc. IEEE Micro Electro mechanical systems conference (MEMS)*, Salt Lake City, Utah, USA, 20–22 February, 1989, pp. 133–138.

[2.50] K. L. Yang, D. Wolcoxen and G. Gimpelson, The effects of post processing techniques and sacrificial layer materials on the formation of free standing polysilicon microstructures, *Proc. IEEE Micro-mechanical systems*, Utah, USA, 20–22 February, 1989, pp. 66–70.

[2.51] H. Guckel, J.J. Sniegowski, T.R. Christenson, S. Mohney and T.F. Kelly, Fabrication of micromechanical devices from polysilicon films with smooth surfaces, *Sensors and Actuators*, 20 (1990) 117–122.

[2.52] P. J. French, B. P. van Drieënhuizen, D. Poenar, J. F. L. Goosen, P. M. Sarro and R. F. Wolffenbuttel, Low-stress polysilicon process compatible with standard device processing, *Sensors VI: Technology, Systems and Applications*, Manchester, UK, 1993, pp. 129–133.

[2.53] H. Guckel, T. Randazzo and D.W. Burns, A simple technique for the determination of mechanical strain in thin films with applications to polysilicon, *J. Appl. Phys.*, 57 (1985) 1671–1675.

[2.54a] M.G. Allen, M. Mehregany, R.T. Howe and S.D. Senturia, Microfabrication structures for the *in-situ* measurement of residual stress, Young's modulus, and ultimate strain of thin films, *Appl. Phys. Lett.*, 51 (1987) 241–243.

[2.54b] B.P. van Drieënhuizen, J.F.L. Goosen, P.J. French and R.F. Wolffenbuttel, Comparison of techniques for measuring both compressive and tensile stress in thin films, *Sensors and Actuators*, A37 (1993) 756–7655.

[2.55] M. Sekimoto, H. Yoshihara and T. Ohkubo, Silicon nitride single-layer x-ray mask, *J. Vac. Sci. Tech.*, 21 (1982) 1017–1021.

[2.56] K. Aratani, P.J. French, P.M. Sarro, D. Poenar, R.F. Wolffenbuttel and S. Middelhoek, Surface micromachined tuneable interferometer array, *Sensors and Actuators*, A43 (1994) 17–23.

[2.57] P.J. French, P.M. Sarro, R. Mallée and R.F. Wolffenbuttel, Optimization of a low stress silicon nitride process for surface micromachining applications *Proceedings Eurosensors*

94, Toulouse, France, 26–28 September 1994.

[2.58] Y.X. Li, Private communication

[2.59] P.J. French and R.F. Wolffenbuttel, Reflow of BPSG for sensor applications, *J. Micromech. Microeng.*, 3, (1993), pp. 1–3.

[2.60] A.C. Adams, *Solid State Technology*, 24 (1981) 178.

[2.61] M. Bartek, P.J. French and R.F. Wolffenbuttel, Planarization in surface micromachining using selective epitaxial growth, *Proceedings Eurosensors 94*, Toulouse, 26–28 September 1994.

[2.62] M. Bartek, Y.X. Li, P. Gennissen, P.J. French and R.F. Wolffenbuttel, Epitaxial layer optimization for smart silicon sensors by placing read-out electronics in SEG silicon, *Proceedings Eurosensors 94*, Toulouse, 26–28 September 1994.

[2.63] D.J. Monk, D.S. Soane and R.T Howe, Sacrificial layer SiO_2 wet etching for micromachining applications, *Proceedings Transducers'91*, San Francisco, USA, pp. 647–650.

[2.64] D.J. Monk, D.S. Soane and R.T Howe,A diffusion reaction model for HF etching of LPCVD phosphosilicate glass sacrificial layer, *Proceedings Solid State Sensor and Actuator*, IEEE Hilton Head, USA, 1992, pp. 47–51.

[2.65] D.J. Monk, D.S. Soane and R.T Howe,Determination of the etching kinetics for the hydrofluoric acid/silicon dioxide system, *J. Electrochem. Soc.*, 140 (1993) 2339–2345.

[2.66] D.J. Monk, D.S. Soane and R.T Howe,Enhanced removal of sacrificial layers for silicon surface micromachining, *Proceedings Transducers'93*, Yokohama, Japan, pp. 280–283.

[2.67] P.R. Scheeper, J.A. Voorthuijzen, W. Olhuis and P. Bergveld, Investigation of attractive forces between silicon nitride microstructures and an oxidized substrate, *Sensors and Actuators*, A30 (1992) 231–239.

[2.68] C.H. Mastrangelo and C.H. Hsu, Mechanical stability and adhesion of microstructures under capillary forces-part I: Basic theory; part II. Experimental, *J. Micromechanical Systems*, 2 (1993) 33–55.

[2.69] G.T. Mulhern, D.S. Soane and R.T Howe,Supercritical carbon dioxide drying of microstructures, *Proceedings Trans-*

ducers'93, Yokohama, Japan, pp. 296–299.

[2.70] H.A.C. Tilmans, Micromechanical sensors using encapsulated built-in resonant strain gauges, PhD thesis, Twente University, The Netherlands, 1993.

[2.71] Y.X. Li, M.R. Wolffenbuttel, P.J. French, M. Laros, P.M Sarro, and R.F. Wolffenbuttel, Reactive ion etching (RIE) techniques for micromachining applications, *Sensors and Actuators*, A41 (1994) 317–323.

[2.72] T.A. Core, W.K. Tsang and S.J. Sherman, Fabrication technology for an integrated surface-micromachined sensor, *Solid State Technology*, Oct. 1993, pp. 39–47.

[2.73] M.R. Wolffenbuttel, Y.X. Li, D. Poenar, P.J. French, P.P.L. Regtien and R.F. Wolffenbuttel, Multilayer membranes for a tactile imaging sensor, using surface micromachining and RIE, *Proceedings Transducers 93*, Yokohama, Japan, 1993, pp. 284–287.

[2.74] B.P. van Drieënhuizen, P.J. French, Y.X. Li, D. Poenar and R.F. Wolffenbuttel, Integrated capacitive rms to dc converter based on surface micromachining technology, *Proceedings Eurosensors 94*, Toulouse, France, 26–28 September 1994

[2.75] E.W. Becker, W. Ehrfeld, P. Hagmann, A. Maner and D. Munchmeyer, Fabrication of microstructures with high aspect ratios and great structural heights by synchrotron radiation lithography, galvanoforming and plastic moulding (LIGA process), *Microelectronic Engineering*, 4 (1986) p. 35.

[2.76] H. Guckel, K.J. Skrobis, T.R. Christensen, J. Klein, S. Han, B. Choi, E.G. Lovel and T.W. Chapman, On the application of deep X-ray lithography with sacrificial layers to sensor and actuator construction (The magnetic micromotor with power take-offs), Tech. Digest, *Transducers 91*, (1991), p. 393.

[2.77] L.A. Field and R.S. Muller, Fusing silicon wafers with low melting temperature glass, *Sensors and Actuators*, A21–23 (1990) 935–938.

[2.78] Y. Kanda, K. Matsuda, C. Murayama and J. Sugaya, The mechanism of field assisted silicon-glass bonding, *Sensors and Actuators*, A21–23 (1990) 939–943.

[2.79] E.F. Cave, Method of making a composite insulator semiconductor wafer, US Patent No. 3,290,760, 1966.

[2.80] M. Esashi, A. Nakano, S. Shiji and H. Hebiguchi, Low temperature silicon-to-silicon anodic bonding with intermediate low melting point glass, *Sensors and Actuators*, A21–23 (1990) 931–934.

[2.81] R.F. Wolffenbuttel and K.D. Wise, Low-temperature silicon wafer-to-wafer bonding using gold at eutectic temperature, *Sensors and Actuators*, A43 (1994) 223–229.

[2.82] T. Hamaguchi, N. Endo, M. Kimura and M. Nakamae, Novel LSI/SOI wafer fabrication using device layer transfer technique, *Proc. IEDM'86*, pp. 688–691.

[2.83] A.D. Brooks, R.P. Donovan and C.A. Hardesty, Low-temperature electrostatic silicon to silicon seals using sputtered borosilicate glass, *J. Electrochemical Soc.*, 119 (1972) 545–546.

[2.84] H.J. Quenzer and W. Beneke, Low temperature silicon wafer bonding, *Sensors and Actuators*, A32 (1992) 340–344.

[2.85] V.M. McNeil, M.J. Novack and M.A. Schmidt, Design and fabrication of thin-film microaccelerometers using wafer bonding, *Proceedings Transducer 93*, Yokohama, pp. 822–825.

[2.86] P.W. Barth, Silicon fusion bonding for fabrication of sensors, actuators and microstructures, *Sensors and Actuators*, A21–23 (1990) 919–926.

[2.87] M. Wale and M. Goodwin, Flip-chip bonding optimizing opto-IC's, *IEEE Circuits Devices*, No. 11 (1992) 25–31.

3

COMPATIBILITY OF PIEZOELECTRIC ZINC OXIDE WITH SILICON IC PROCESSING

3.1 Introduction: piezoelectricity and silicon

In certain materials, a mechanical stress produces an electric charge and an applied electric field generates a mechanical strain. Such materials are called piezoelectric. Some naturally found materials exhibit this effect (quartz, Rochelle salt), and numerous synthetic piezoelectric materials have been developed. Piezoelectric materials possess a crystal structure that lacks a center of symmetry and have a dominating ionic bonding. The piezoelectric effect is used in sensor devices for the transduction of mechanical energy to electrical energy and vice versa.

As mentioned in chapter 1, silicon is a non-piezoelectric material (the silicon crystal is symmetric and the bonding between the atoms is covalent). The realization of silicon-based piezoelectric transducers requires the deposition of a piezoelectric overlay.

This piezoelectric overlay has to satisfy demands with respect to the electro-mechanical coupling factor, the acoustic velocity, the elastic properties, the electrical properties (such as conductivity), and the ease of fabrication. For silicon device applications, the fabrication processes have to be compatible with standard IC processing (bipolar or CMOS).

Several overlay materials have been investigated in the last 25 years. In section 3.2, some of these materials are reviewed. Zinc oxide is the most widely used piezoelectric overlay on silicon. The IC-compatible fabrication processes do not only concern the growth but also the further processing of the zinc oxide such as patterning and the fabrication of metal electrodes on top of the zinc oxide. The zinc-oxide process module that was developed at the Delft University of Technology, DIMES, is discussed in section 3.3.

Some applications of zinc oxide on silicon devices are discussed in section 3.4: a thin-film resonator, a force sensor, a tactile-sensor array, an acousto-optic light deflector, and Lamb wave sensor systems. In the final section the conclusions are given.

3.2 Thin film piezoelectric materials on silicon

In the field of acoustic wave devices, a strong interest in piezoelectric thin films was developed in the early seventies. Layered surface acoustic wave (SAW) filters were realized on glass, sapphire and silicon substrates. Research was focussed on the fabrication of piezoelectric cadmium sulfide films (CdS) and in particular zinc oxide (ZnO) and aluminum nitride films (AlN) [3.1–3.6]. At the same time, the deposition and polarizing of polymers on non-piezoelectric substrates were investigated. Piezoelectric polyvinylidene fluoride (PVDF) is an attractive material for transducer applications (e.g. tactile sensors and transducers for non-destructive testing). The properties of some materials are compared in Table 3.1.

PVDF films are produced in the form of a sheet (typical thickness between 10 and 100 μm) and piezoelectricity is obtained after polarization (by applying a DC voltage over the two sides of the sheet). For this purpose, both sides of the sheet are provided with a thin metal film. The sheet is glued to the non-piezoelectric substrate. The metal film on the sheet can be patterned by using photolithographic techniques.

PVDF films can also be produced by spinning. Polarization is obtained by placing the substrate in a strong electric field, after the deposition of the PVDF. The deposition is an IC-compatible

Table 3.1 Properties of some piezoelectric overlay materials

Material	Fabrication	Electro-mechanical coupling constant	Acoustic-wave velocity $[ms^{-1}]$	density $[kgm^{-3}]$
PVDF	sheet/ spinning	0.17	2200	1800
AlN	sputter	0.17	5600	3260
CdS	sputter	0.15	1800	4820
ZnO	sputter	0.28	2500	5400

process which is an advantage when aiming at silicon integration [3.7].

In surface acoustic wave applications the use of PVDF transducers is not attractive. The low Young's modulus of PVDF (typically 100× lower than for the other given materials) causes high propagation losses. For acoustic imaging, this low elastic modulus is no obstacle and silicon-PVDF transducer arrays have been fabricated [3.8].

The other three materials of Table 3.1 (AlN, CdS, ZnO) are mainly fabricated by sputtering techniques. The interest in CdS has faded due to problems in reproducible fabrication and because of its toxic character. AlN and ZnO are still used in many transducers. The coupling factor of AlN is somewhat lower than that of ZnO. The surface acoustic wave (SAW) velocities of the two materials are far apart: for ZnO it is fairly low (about 2500 ms^{-1}), while the velocity in AlN is rather high (5600 ms^{-1}). In this sense the two materials can be seen as complementary. It depends on the application, if a low or a high velocity is advantageous. The reproducible fabrication of AlN seems to give a few more problems than the fabrication of ZnO. With respect to silicon integrated piezoelectric transducers, ZnO is the most widely used material. The fabrication and use of this material will be discussed in detail in the next section.

3.3 IC-compatible ZnO fabrication

For the realization and further processing of a piezoelectric zinc oxide film on the silicon substrate, extra process steps should be added to the standard IC processes. As mentioned in chapter 1, the integration of these steps in an IC fabrication process is hampered by the restrictions imposed by the process. The process steps and material properties are subjected to tight control in order to maintain the standards necessary to produce functioning electronic devices. In order to prevent the interference of the IC process, the sensors are designed such that all, or most, of the extra sensor processes take place after the completion of the standard IC-process. Both the Delft and Twente Universities of Technology have succeeded in using this set-up [3.9–3.10]. In Delft, a general fabrication strategy is used in which the sensor process steps are placed in 'process modules'. Such modules are added to the well-established IC-process. The zinc oxide module (ZnO-module) comprises the growth and the further processing of the zinc oxide film, e.g. patterning of the zinc oxide film and fabrication of metal electrodes on top of the zinc oxide.

For low-frequency applications extra measures have to be taken to avoid current leakage through the zinc oxide. This is briefly reviewed in section 3.3.1.

As an example of the use and fabrication of piezoelectric zinc oxide films on silicon devices, the fabrication sequence of a bipolar monolithic acoustic oscillator based on Lamb waves is described in section 3.3.2. For the fabrication of these devices, bipolar IC-technology, zinc oxide technology and micromachining technology were combined. The micromachining technology used for this device is not further discussed.

The zinc oxide fabrication, characterization, and further processing is described in section 3.3.3.

3.3.1 Low-frequency applications

At low frequencies, the build-up of an electric field in the zinc oxide film is hindered by a relatively high conductivity caused by excessive zinc [3.11]. At high frequencies (above 1 MHz), the conductivity problem does not occur.

For low-frequency and static applications, some methods have been presented to increase the resistivity of the zinc oxide [3.12]. These methods are the doping of the film by lithium, encapsulation of the zinc oxide between two dielectric layers (e.g. silicon dioxide), using a MOS structure, or using an air-gap structure. It is clear that it imposes some restrictions to the design of the device. Therefore, the usefulness of these methods has to be reconsidered for each application.

The prevention of current leakage over the zinc oxide layer is not further discussed in this chapter.

3.3.2 *Device fabrication sequence*

A monolithic Lamb wave oscillator consists of an acoustic delay line implemented in a ZnOAl-SiO_2-Si layered structure which is monolithically integrated on one chip with the bipolar amplifier circuitry. The acoustic waves are sensitive to modulations taking place at the surface of the substrate in which they propagate. This specific sensor is used for the determination of the physical properties of liquids (such as density and viscosity), but also for chemical sensing in both gases and liquids. The operation of this sensor is further discussed in section 3.4.5.

The complete fabrication sequence of the Lamb-wave sensor is described with the aid of Fig. 3.1. For each step the process module is given (the zinc oxide processes are denoted as the ZnO-module, the bulk-micromachining processes as the BMM1-module). The process steps of these two modules have been placed behind the standard bipolar IC-process, as can be seen in the figure.

In this way, the interference of the extra process steps with the standard process is prevented.

3.3.3 *Zinc oxide processing*

Advantages of piezoelectric zinc oxide films are the relatively low deposition temperature (<450°C), the high piezoelectric coupling coefficient and the excellent bonding on various substrate materials, in particular silicon, silicon dioxide, silicon nitride, and metals such as gold and aluminum. The combination of the silicon substrate and zinc oxide provides the possibility of monolithic integration of acoustic elements with electronic circuitry. Examples of monolithic acousto-electronic devices are given in section 3.4.

Zinc oxide also has some disadvantages. It is amphoteric and

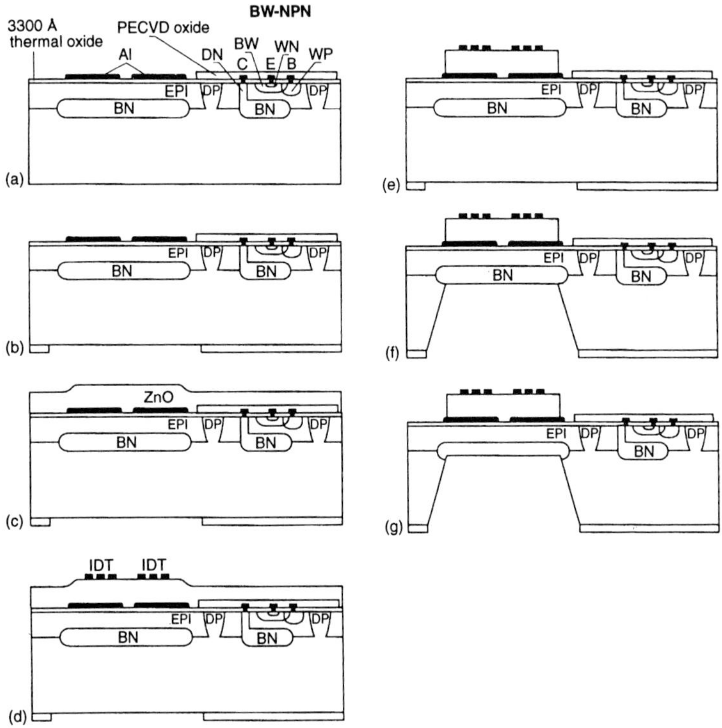

Figure 3.1. *Fabrication sequence of the monolithic Lamb-wave sensor. (a) Standard IC-process. The starting material is double-side polished p-type <100>-cut silicon, wafer diameter 100 mm and thickness 525 μm. The wafers undergo the standard IC-process, including the interconnect metalization (0.6 μm) and a PECVD oxide passivation (0.8 μm). The oxide is removed from the propagation path. (b) BMM1-module. Back deposition and patterning (plasma-etched) of PECVD nitride (360 nm), used as masking layer during the bulk-etch process. (c) ZNO-module. A piezoelectric zinc oxide layer is deposited on the front of the wafer. (d) ZNO-module. Deposition and patterning (wet-etch) of 250 nm aluminum (100% Al, vacuum evaporation). (e) ZNO-module. The zinc oxide above the electronic circuitry is removed (wet-etch). (f) BMM1-module. KOH bulk etching of silicon (plate fabrication). The standard silicon n-epilayer thickness of the DIMES-01 process is 4 μm. To increase the plate thickness, a BN implantation is used. The silicon is etched using an electro-chemical etch-stop at the BN-substrate junction, leaving a 7.5 μm thick plate. (g) BMM1-module. If required, the plate can be thinned by plasma etching of the back.*

chemically not very stable. When used in devices, it has to be protected against the environment by gas-tight packaging (housing filled with an inert gas). Furthermore, the fabrication of reproducible high-quality zinc oxide requires an extremely well-controlled deposition process. Also, zinc oxide films can show high stress due to the growth process. Knowledge of the influence of the deposition parameters and crystal orientation of the surface of the substrate material on the residual stress is of crucial importance.

As mentioned, the zinc oxide module not only incorporates the deposition and patterning of piezoelectric zinc oxide films, but also the deposition and patterning of aluminum on top of zinc oxide. Some special cleaning fluids and wet etchants which do not attack zinc oxide are used. The first step of the module consists of the deposition of the zinc oxide on the fully processed silicon wafer. The typical zinc oxide layer thickness for Lamb-wave applications is between 0.7 and 2 μm. On top of the zinc oxide, an aluminum film is deposited (250 nm thick) in which the interdigital patterns are etched. The last step consists of the removal of the zinc oxide from the electronic circuitry.

The steps used in zinc oxide processing are described in the following sections.

(a) Deposition of ZnO

High-quality piezoelectric zinc oxide is deposited mainly by sputtering techniques, e.g. triode sputtering, DC and RF magnetron sputtering. The process temperature is between 150 and 450°C and sputter rates of up to 20 μmhr^{-1} are feasible. Sputtered zinc oxide layers are dense, highly oriented, and show a high degree of surface flatness. Zinc oxide has a hexagonal structure, class 6-mm (Fig. 3.2). The crystals of the polycrystaline layer show a preferential growth of their c-axis perpendicular to the substrate surface. Highly oriented films show a high piezoelectric coupling factor. High-quality zinc oxide layers of thicknesses between 0.1 and 48 μm have been reported [3.13–3.23].

The zinc oxide layers reported in this work were fabricated in a Varian 3119 R&D DC sputter-coater which uses an S-gun magnetron (see Fig. 3.3).

A set of typical sputtering conditions is given in Table 3.2.

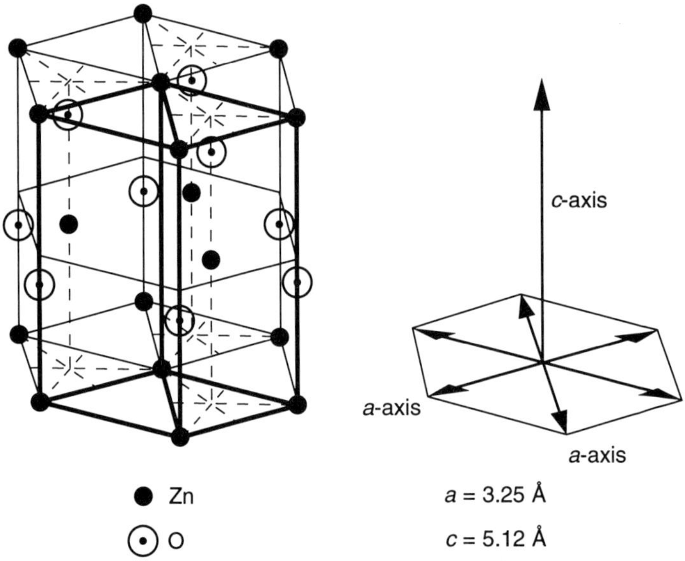

Figure 3.2. *The structure of ZnO.*

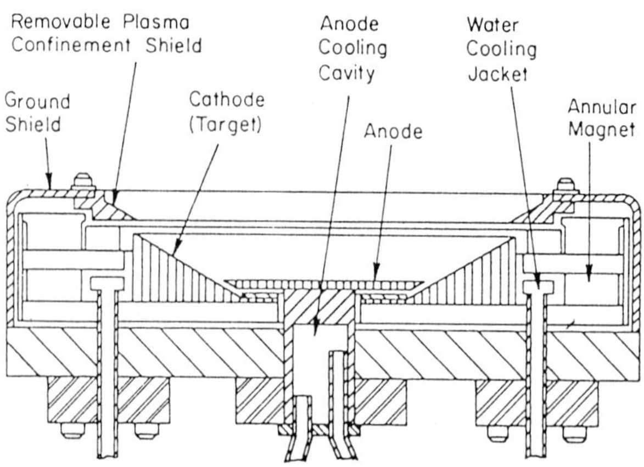

Figure 3.3. *Cross-section of the S-gun sputter electrodes with B field and plasma ring.*

Table 3.2 Typical ZnO deposition conditions

Target material	Zn (99.999%)
Sputter gas	O_2
Substrate temperature	375 °C
Target–substrate spacing	11 cm
Oxygen pressure	0.9 kPa
DC sputter power	1.5 kW (500 V, 3 A)
Sputter rate	100 nmmin^{-1}

(b) Film characterization

The quality of piezoelectric zinc oxide films has been a subject in the literature for over 25 years. Sputter deposition of zinc oxide shows superior properties over other deposition methods, but the growth and quality remains a subject of concern, as recent publications bear witness [3.24–3.25].

For the characterization of the piezoelectric zinc oxide films, several methods can be used: scanning electron microscopy (SEM), X-ray diffraction ($2\theta_i$-scan, rocking curve), and surface profiling (uniformity and surface roughness).

Scanning electron microscope (SEM) The morphology of the zinc oxide films is visualized by using scanning electron microscopy (SEM). SEM gives information about crystal orientation, grain size, and porosity.

The SEM-picture in Fig. 3.4 shows a layer fabricated by using the typical growth conditions. The zinc oxide layer is polycrystaline with the c-axis oriented perpendicular to the substrate surface. The thickness of the layer shown is about 10 μm, a clear difference between the morphology of the first microns of zinc oxide and the zinc oxide above the first microns can be seen. In the first microns, the zinc oxide is denser and the polycrystaline structure cannot be seen clearly. The difference in structure affects the etch rate of the zinc oxide, as discussed below.

X-ray diffraction Zinc oxide films grown on Al-SiO_2-Si and Al-Si layered structures were examined by X-ray diffraction using a Siemens D500 diffractometer with diffracted-beam graphite monochromator (Cu-Kα radiation). Highly oriented zinc oxide layers

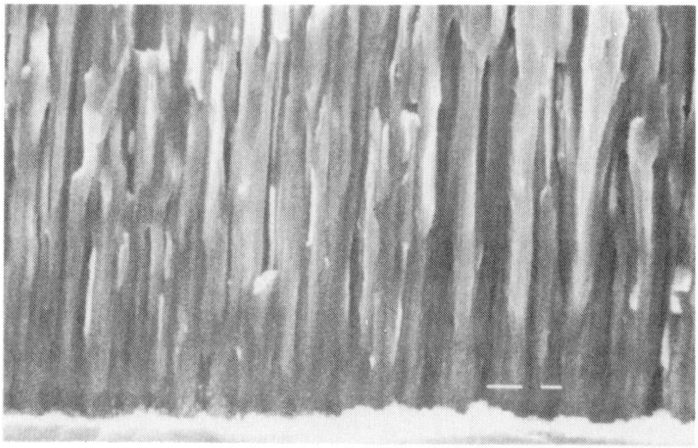

Figure 3.4. *Cross-view SEM picture of a zinc oxide layer fabricated by using the standard growth conditions.*

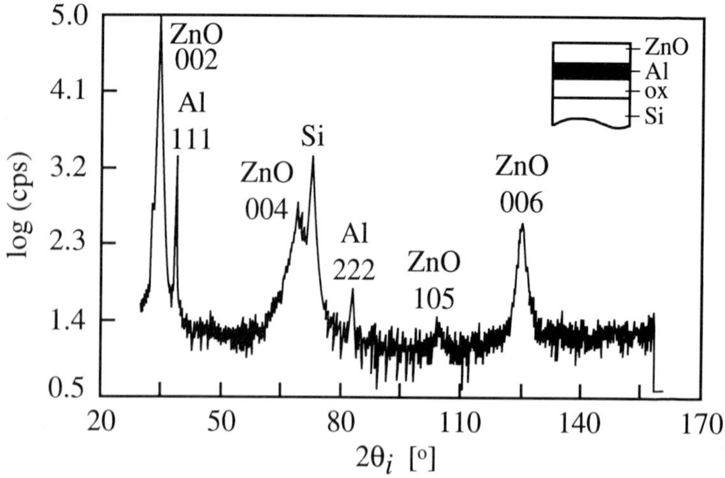

Figure 3.5. *$2\theta_i$-scan of a 1.2 μm thick ZnO layer on an Al-SiO_2-Si layered structure (the aluminum has a strong (111) orientation).*

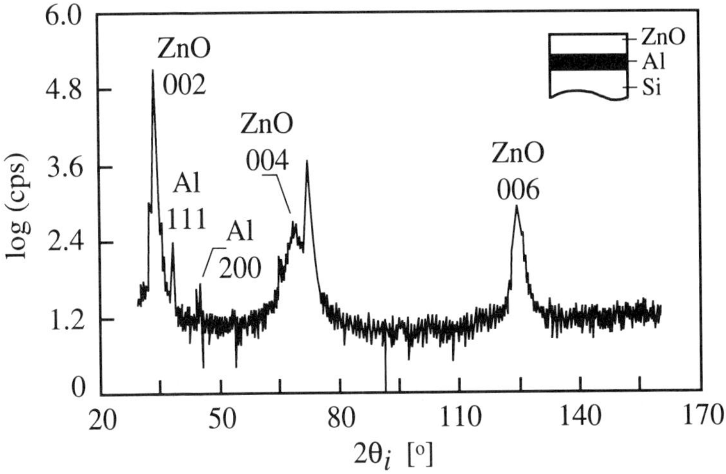

Figure 3.6. *$2\theta_i$-scan of a 1.2 μm thick ZnO layer on an Al-Si layered structure (the aluminum has a (111) orientation, but the (200) peak is also present).*

show a strong reflection intensity from the (002) planes (and, therefore, also strong (004) and (006) peak intensities), and low intensities from other zinc oxide planes (such as (102), (103), (104), (105)). Figure 3.5 shows a typical $2\theta_i$-scan of a 1.2 μm thick zinc oxide layer grown using the typical growth parameters, on an Al-SiO_2-Si layered substrate. The peak intensity is given in counts per second (cps). The (002), (004), and (006) peaks are clearly present, and of the other possible peaks only the (105) shows some intensity. The aluminum shows a strong (111) orientation, no other than the (111) and (222) peak were found.

In Fig. 3.6 a similar scan is obtained for a ZnO-Al-SiO_2 layered structure. The aluminum was grown directly on the silicon substrate, and shows some intensity for the Al(200) peak. Comparison of this figure with Fig. 3.7 shows the influence of the orientation of the aluminum layer on the orientation of the subsequently grown zinc oxide layer. The aluminum in Fig. 3.7 presents not only the (111) and (200) peaks, but also (311), (331), (420), and (422). The

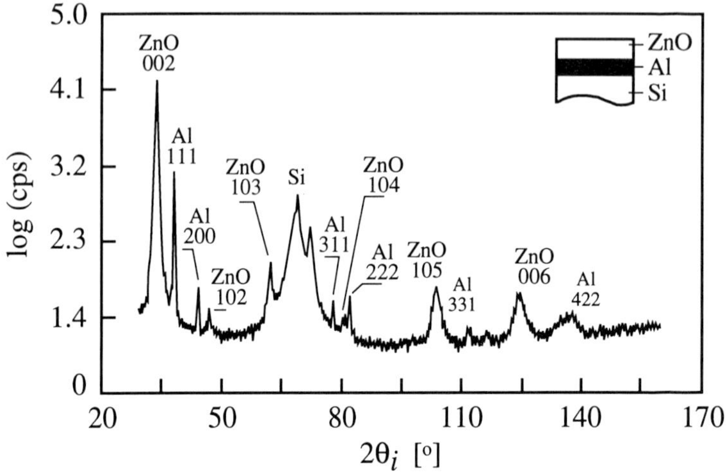

Figure 3.7. *$2\theta_i$-scan of a 1.2 μm thick ZnO layer on an Al-Si layered structure (both the aluminum and the zinc oxide show several peaks).*

Table 3.3 *c*-axis distribution of zinc oxide films grown on different substrates

Substrate material	*c*-axis distribution
Si	±15 °
$Al_{(111)}$-SiO_2-Si	±6 °
$Al_{(111)}$-Si	±3 °
$Al_{(low(111)orientation)}$-Si	±8 °

zinc oxide shows in addition to the (002), (004), and the (006) peaks, intensities for the (102), (103), (104), and (105) peaks. The (piezoelectric) quality of this zinc oxide layer is low. It is of crucial importance, therefore, that highly (111) textured aluminum is used. A precise description of the growth mechanisms is not available, but it is assumed that growing zinc oxide on substrates of (closely) matching crystal symmetry improves its crystallinity [3.19–3.20,3.22–3.23].

By measuring X-ray rocking curves, the distribution of the *c*-axis orientation is determined [3.26]. In Table 3.3, the degree of crystal orientation for ZnO layers grown on different substrate materials but under equal conditions is given. The substrate material has a clear influence on the c-axis distribution.

Uniformity and surface roughness of ZnO layers The uniformity and surface roughness measurements were performed using an alpha step 200 (Tencor Instruments). The uniformity is measured over a 100 mm diameter wafer (substrate material: p-type <100>-cut silicon, standard deposition parameters, zinc oxide layer thickness 1.5 μm). The non-uniformity within a 10 mm diameter circle amounts to ±1%. The surface roughness of a 1.5 μm thick piezoelectric zinc oxide layers was about 4 nm.

(c) Process steps

ZnO is a reactive material and it is attacked by standard etch solutions such as aluminum etch. Also standard cleaning procedures can damage the ZnO layer, and ZnO could cause contamination of the process chambers. Therefore, different process steps have to be used once the ZnO layer has been deposited on the wafer [3.27]. The processes are discussed in the following sections.

Cleaning After the deposition of the zinc oxide the wafers have to be cleaned before further processing. The standard cleaning fluids (fuming 100% nitric acid, water) damage the ZnO layer and cannot be used. Organic solvents, on the other hand, do not affect the zinc oxide. A cleaning cycle containing acetone, toluene and dimethylsulfoxide (DMSO) was used (see Table 3.4). During cleaning in these fluids, ultrasonic agitation was applied.

Etching Zinc oxide is attacked by all common acids and bases. Because of the crystaline structure of zinc oxide, the etch rate of many etchants is direction-dependent. Grain size and porosity of the zinc oxide layer also affect the etch rate [3.14]. It should be noted that the structure of the zinc oxide at the start of growth, approximately the first few microns, is different from the structure above the first microns (as is shown in Fig. 3.4). For slow Lamb-wave devices, only thin zinc oxide films are required, typically 0.7–2 μm thick. The etchants used to pattern the zinc oxide are given in

Table 3.4 Three-step cleaning cycle for zinc oxide

1	Acetone, 5 min; DI water rinse, 1 min; dry.
2	Toluene, 5 min; DI water rinse, 1 min; dry.
3	DMSO (dimethylsulfoxide), 5 min; DI water rinse, 1 min; dry.

Table 3.5 Etchants for etching of zinc oxide films

	Etchant	Etch rate series 1 [μmmin^{-1}]	Etch rate series 2 [μmmin^{-1}]
a	H_3PO_4, 10 ml HAc, 10 ml H_2O, 100 ml	1.5	0.5
b	H_3PO_4, 1 ml HAc, 100 ml H_2O, 100 ml	0.8	0.2

Table 3.5. The etch temperature was 20°C and the etchants were stirred. HPR204 photoresist was used as a masking layer.

The test samples of series 1 were oxidized silicon substrates covered with 1 μm thick zinc oxide. The samples of series 2 were oxidized silicon covered by a 600 nm thick sputtered aluminum layer (containing 1% silicon), the zinc oxide layers were between 0.8 and 1.7 μm thick. The above etch rates all relate to etching parallel to the *c*-axis of the zinc oxide crystal. The etchants show a lower etch rate for zinc oxide grown on aluminum, which indicates that the quality of zinc oxide grown on aluminum is higher than that grown on silicon dioxide. For the fabrication processes reported here, etchant #a was used for the fast etching of thick ZnO layers, whereas etchant #b was used for slower etching. The etch rate of these mixtures varies approximately by 30% with a temperature

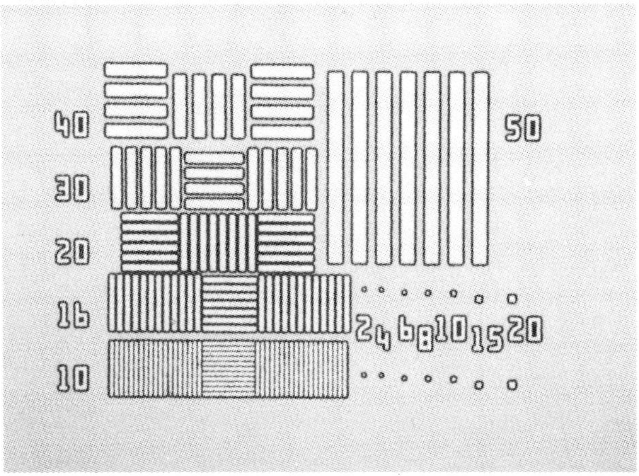

Figure 3.8. *Photograph showing the undercutting of a 1 μm thick zinc oxide film after wet-etching. The dark areas represent zinc oxide, in the light areas the zinc oxide is etched.*

change of 5°C. The shelf lives of solutions #a and #b are at least one week.

The undercutting of a 1 μm thick zinc oxide layer is about 3 μm for etchant #a, and 5 μm for etchant #b. The lateral etching is faster than the axial etching, but in this application the undercutting of the zinc oxide is not critical. By using a dry-etching process, undercutting can be minimized [3.28].

In Fig. 3.8, the undercutting of a 1 μm thick zinc oxide using etchant #b is shown. The dark areas represent the zinc oxide, in the light areas the zinc oxide is etched. In the test blocks, the gaps and lines were designed with equal width (the values are given in μm in the photograph). The photograph shows that in the 10 μm test block the lines are almost totally etched, indicating 5 μm undercutting.

Fabrication of aluminum patterns on top of zinc oxide layers For the fabrication of aluminum patterns on top of zinc oxide, an aluminum layer on top of the zinc oxide is vacuum evaporated. This layer should be kept as thin as possible, since thick aluminum patterns induce more reflections of the acoustic waves,

Table 3.6 Etchants for the etching of aluminum patterns on top of zinc oxide

Etchant	Etch rate [μmmin^{-1}]
1 g KOH, 10 g $K_3Fe(CN)_6$, 100 ml H_2O	0.5
1 g KOH, 10 g $K_3Fe(CN)_6$, 600 ml H_2O	0.2

thus affecting the device performance. A minimum thickness of 200 nm is required to ensure good electrical conductivity.

For the etching of the aluminum, a solution of KOH and $K_3Fe(CN)_6$ in water can be used (see Table 3.6). For the thin aluminum layers the slow etch solution was used. The etch temperature was 20°C and the etchants were stirred. HPR204 photoresist was used as a masking layer.

The undercutting of the aluminum depends only on its thickness; the axial and lateral etch rates are equal.

(d) Aspects of mechanical prestress in zinc oxide

The main contributor to stress in the ZnO-Al-SiO_2-Si layered structure is the zinc oxide layer. The slow Lamb-wave phase velocity increases significantly when in-plane stress (or prestress) occurs. An excess of stress could increase the velocity such that it exceeds the sound velocity of the loading liquid, thus causing dramatic propagation losses. In addition, stressed zinc oxide will show relaxation behavior in time and cause aging effects which change the specifications. When zinc oxide films are used in a thin layered structure such as the slow Lamb-wave device, the stress causes buckling, which makes the plate very fragile and unusable for sensor purposes.

For these reasons, the influence of the growth parameters on the residual stress in zinc oxide was investigated. X-ray diffraction analysis was used to determine the amount of stress occurring in the layered structure.

In addition, the usefulness of a stress-relaxation process to decrease the zinc oxide stress after deposition was examined.

Stress in zinc oxide films Stress in deposited thin films usually contains a thermal component and an intrinsic component. The thermal component is caused by the difference between the film deposition temperature and the device operating temperature. The thermal expansion coefficient of zinc oxide (a-axis: 4×10^{-6} K^{-1}) is larger than that of silicon (2.33×10^{-6} K^{-1}), and thus there is a tensile in-plane stress in the zinc oxide due to the cooling after deposition.

The intrinsic stress is caused by imperfections which occur in the crystals during film growth. The intrinsic stress in zinc oxide is compressive and its magnitude can be much larger than the (tensile) thermal component, thus giving rise to a residual compressive stress and strain. It has been demonstrated that the growth parameters such as temperature, pressure, power, gas mixture, and sputter set-up can have influence on the intrinsic stress [3.24,3.29–3.33].

The intrinsic stress in films with a high melting point, but which have been deposited at a low temperature, is large. At low deposition temperatures, the adsorbing atoms have too little energy to arrange themselves in their lowest energy state and intrinsic stresses are built up. However, if the deposition temperature is too high, other negative effects can take place, such as large grain growth.

At high gas pressure, the particles arriving at the substrate have lower energy due to the greater number of collisions before deposition, and arrangement becomes more difficult. At very low pressure, the energy of the adsorbed atoms can become too high, which causes a bombardment of the substrate. This hampers the formation of the film and causes intrinsic stress.

A higher deposition rate (because of a higher sputter power) results in higher particle energy at the substrate, thus yielding a smoother arranging of the adsorbed atoms. If the rate is too high, the adsorbed atoms might be encapsulated by succeeding atoms before having found their lowest state of energy, which again results in higher intrinsic stress.

Several zinc oxide layers were grown on Si and Al-Si wafers (the aluminum contained 1% silicon and had a thickness of 0.6 μm). The stress in the zinc oxide layer was determined by measuring the (006) peak shift using X-ray diffraction [3.26]. The zinc oxide layers

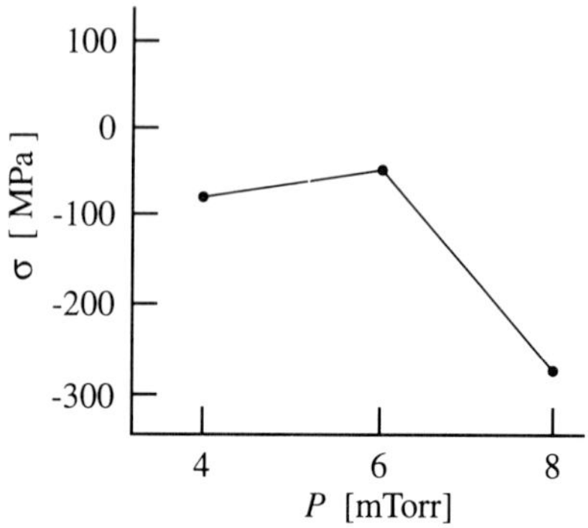

Figure 3.9. *Stress in the ZnO film versus the oxygen pressure (sputter temperature 375°C, sputter current 2.8 A).*

grown directly on silicon showed a much larger peak shift than the zinc oxide layers grown on aluminum-covered silicon. Apparently, the number of defects in zinc oxide grown on aluminum is less than in zinc oxide grown on silicon. The difference in in-plane lattice constants might be an explanation; $a_{\mathrm{ZnO}} = 3.18$ Å, $a_{\mathrm{Si}} = 5.43$ Å, $a_{\mathrm{Al}} = 4.04$ Å. The lattice constant of ZnO matches better with aluminum than with silicon.

The practical influence of pressure, temperature and sputter rate (power) on the stress in the zinc oxide was investigated for ZnO-Al-Si layered substrates. The results are shown in Figs 3.9, 3.10 and 3.11, respectively.

In the range of 4 to 8 mTorr, a pressure of 6 mTorr (100% O_2, temperature 375°C, sputter current 2.8 A) gave the lowest (compressional) stress (Fig. 3.9). Increasing the temperature from 225 to 375°C (6 mTorr, 2.8 A) lowers the residual stress, from −300 to −100 MPa (Fig. 3.10). Increasing the sputter current from 2.4 to 3.1 A (6 mTorr, 375°C) also shows a decreasing compressional stress. At about 3.0 A the stress inverses from compressional to tensile (Fig. 3.11). The compressive intrinsic stress is now of the same order of magnitude as the tensile thermal stress, resulting in a zero residual stress.

Nonetheless, the zinc oxide films deposited on the actual process wafers exhibited higher (compressive) stress levels. The fact that

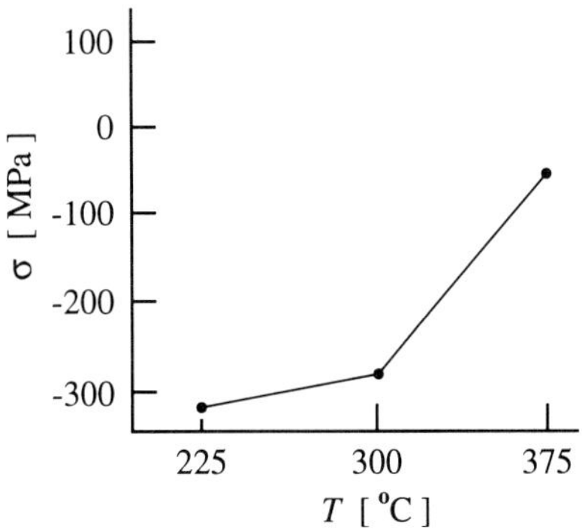

Figure 3.10. *Stress in the ZnO film versus sputter temperature (pressure 6 mTorr, sputter current 2.8 A).*

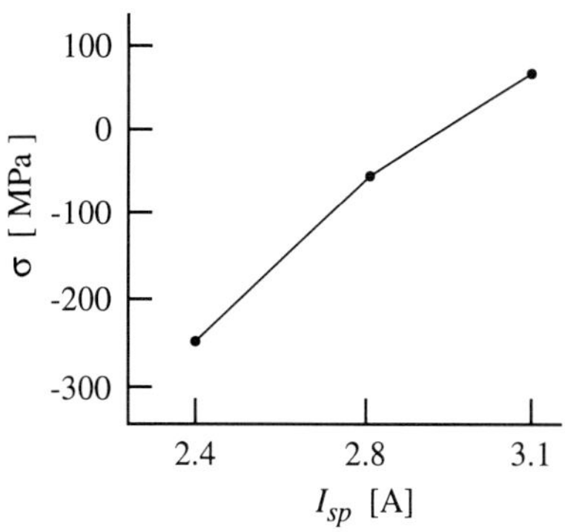

Figure 3.11. *Stress in the ZnO film versus the sputter current (pressure 6 mTorr, sputter temperature 375°C).*

these wafers are only locally covered with aluminum strongly influences the stress.

Stress relaxation Post-deposition relaxation of stress in zinc oxide is possible through annealing. Through the heat treatment boundary defects are eliminated and the *c*-axis orientation of the polycrystaline zinc oxide is improved [3.31,3.34–3.38].

Both ZnO-Si and ZnO-Al-Si layered structures were annealed at

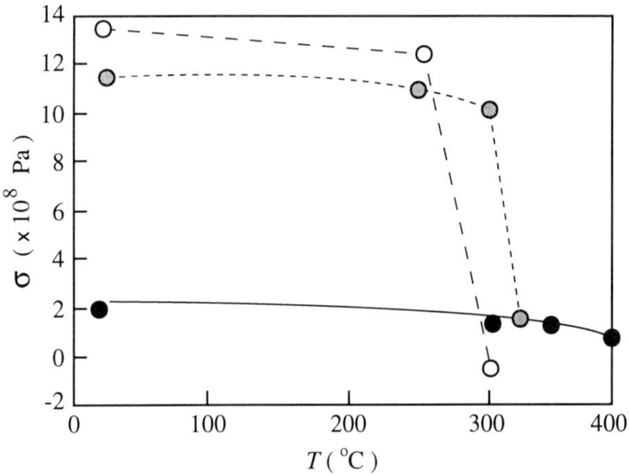

Figure 3.12. *Zinc oxide stress before and after annealing treatment (anneal time of 960 min per measurement). The white and gray dots are from a ZnO-Si layered structure; the black dots are from a ZnO-Al-Si structure.*

temperatures between 280°C and 400°C, for periods up to 16 hours. Annealing was performed in vacuum, and the samples were cooled quickly by placing them on a cold-block (in air). In Fig. 3.12, the results of the annealing is shown for ZnO-Si and ZnO-Al-Si layered samples. The anneal time of each measurement was 960 minutes.

The level of stress was substantially decreased for ZnO-Si layered samples at temperatures above 260°C. For the ZnO-Al-Si layered samples the initial stress was much lower and only at high annealing temperatures (300–350°C) some relaxation was observed. However, the acoustic performance of Lamb-wave devices annealed at 350°C decreases dramatically, possibly due to aluminum diffusion into the zinc oxide.

3.4 Examples of zinc-oxide-on-silicon devices

In this section, some examples of ZnO-based transducers are reviewed. The examples are all different in their application, while

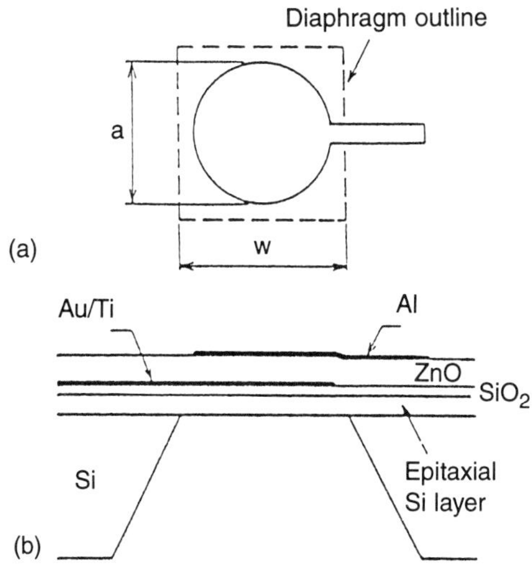

Figure 3.13. *Schematic of the Al-ZnO-Au/Ti-SiO_2 layered resonator. (a) Top view; (b) cross-section (after Miyasaka et al. [3.39]).*

the technology used is similar: a combination of a zinc oxide transducer and electronic circuitry.

The monolithic integration has several, divergent, advantages which are mentioned in the description of the examples.

3.4.1 Thin-film resonator

Miyasaka *et al.* have developed a thin film resonator for an on-chip voltage-controlled oscillator [3.39]. A high degree of temperature compensation was reached by selecting the correct layer thicknesses in the ZnO-SiO_2-Si layered structure.

The schematic of the ZnO-based resonator is shown in Fig. 3.13. By etching the bulk silicon a membrane is fabricated. The membrane consists of a Al-ZnO-Au/Ti-SiO_2-Si layered structure and has a diameter of about 400 μm. The layer thicknesses are 200 nm Al, 10 μm ZnO, 200 nm Au/Ti, 3 μm SiO_2, 7 μm Si.

The resonator was placed in a voltage-controlled oscillator circuit as shown in Fig. 3.14. The operating frequency was about 160 MHz and the temperature coefficient was only 4.5 ppm°C^{-1}.

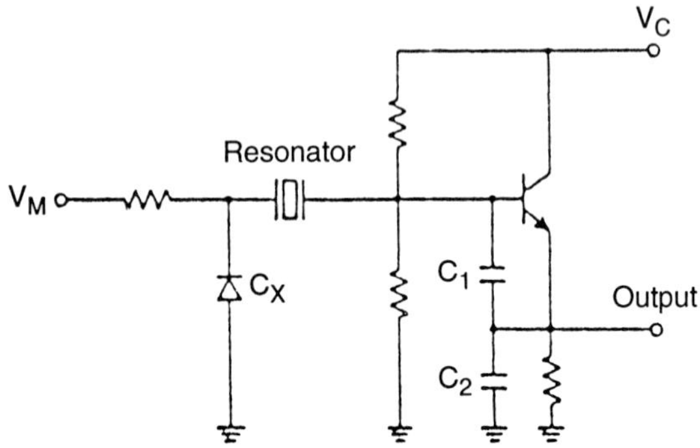

Figure 3.14. *Circuit diagram of the voltage-controlled oscillator (after Miyasaka et al. [3.39]).*

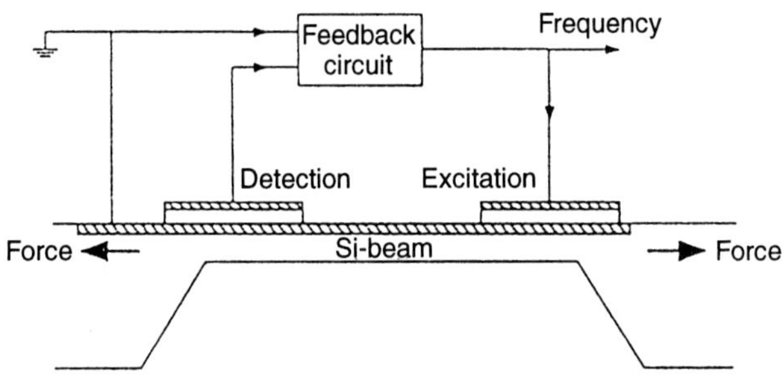

Figure 3.15. *Schematic of the silicon beam oscillator used as a force sensor (after van Mullum [3.12])* .

3.4.2 Force sensor

For several years now, Twente University (The Netherlands) has been working in the field of silicon resonator-beam sensors [3.10–3.11]. A force sensor is one of the applications. Besides the possibility of integration of electronics with the sensor, the silicon implementation allows for very accurate definition of the geometry of the resonating beam by using photolithographic techniques and micromachining technology (<5 μm). In addition, the elastic

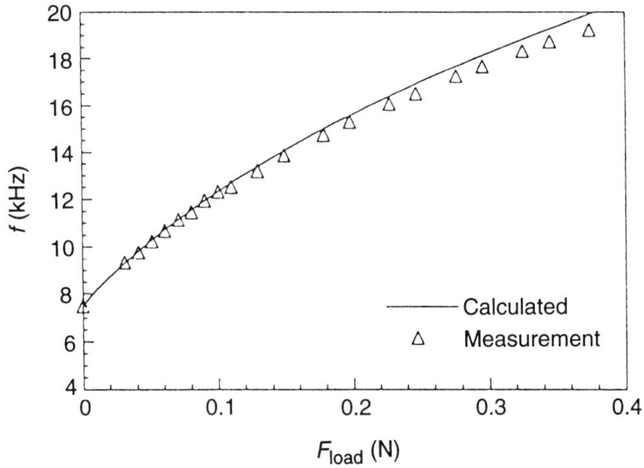

Figure 3.16. *Measured and calculated frequency response of the resonator beam loaded by different forces (after van Mullum [3.12]).*

part (silicon beam) and the transduction material (ZnO) can be optimized separately. A disadvantage is the higher temperature sensitivity of the ZnO-Si transducer compared to (special-cut) quartz transducers.

A drawing of the two-port sensor system is given in Fig. 3.15. The working principle of the force sensor is the shift of the resonance frequency of a silicon beam caused by an external axial force. The beam is brought into resonance by using zinc oxide transducers. In Fig. 3.16 a typical frequency response of the force sensor is shown.

3.4.3 Tactile-sensor array

A silicon tactile sensor-array based on zinc oxide was published by Polla *et al.* [3.40].

The array consists of 1 μm thick sputtered ZnO piezoelectric capacitors in an 8×8 matrix with each sensing element measuring 70×70 μm^2 and spaced with a pitch of 70 μm.

The transducer layout is presented in Fig. 3.17 and the response to a force (mass) is given in Fig. 3.18. The main advantages of the silicon implementation are found in the direct signal coupling of transducer to electronics and the high density of the arrays.

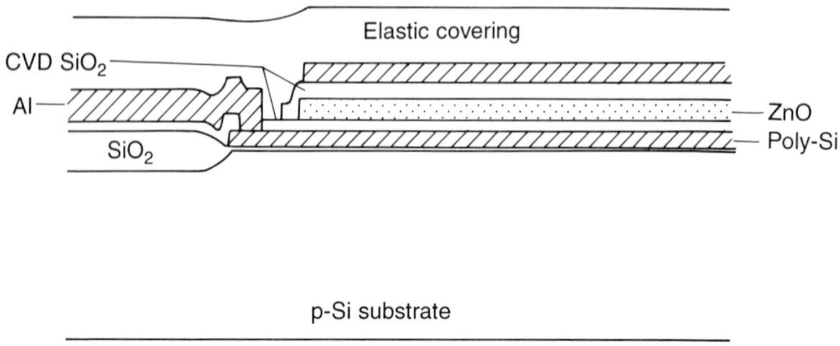

Figure 3.17. *Cross-section of an individual tactile-array element (after Polla et al. [3.40]).*

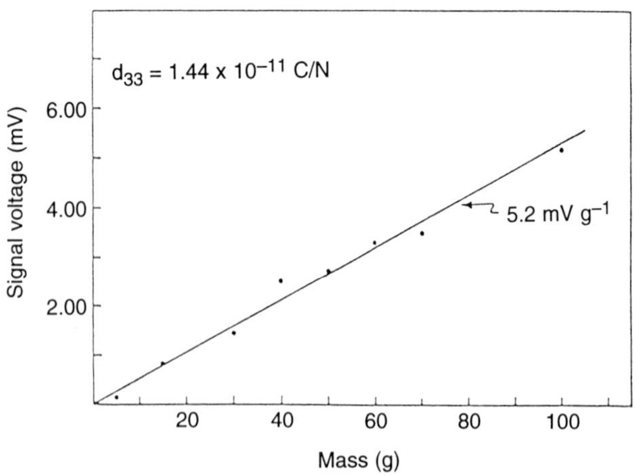

Figure 3.18. *Voltage output as a function of applied nominal load for an individual sensor element (after Polla et al. [3.40]).*

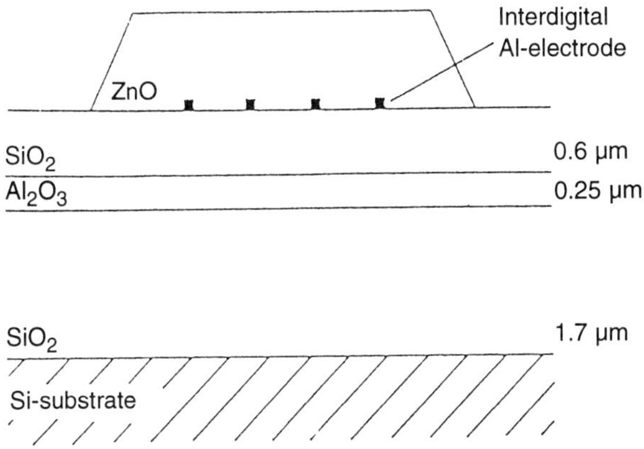

Figure 3.19. *Cross-section of the wave-generating transducer of the acousto-optic deflector (after Benoist et al. [3.42]).*

3.4.4 Deflector in optical waveguides

In integrated optics, surface acoustic waves are used as an acousto-optic Bragg deflector. In a Bragg deflector, an incoming light beam is diffracted by twice the Bragg angle given by $\theta_B = \arcsin(\lambda/2n_{\text{eff}}p)$, where n_{eff} is the effective refractive index of the waveguide structure, and λ and p are the optical and acoustic wavelength, respectively [3.41]. The acoustic wave propagates through the optical waveguide material and produces a strain pattern. This strain affects the refractive index of the wave guide material and an optical grating is realized.

Benoist *et al.* [3.42] presented a deflector in which surface acoustic waves are generated by a ZnO-SiO_2-Al_2O_3-Si layered structure (Fig. 3.19). The aluminum oxide is the optical waveguide material. In Fig. 3.20 the top view of the silicon integrated Bragg deflector is shown.

3.4.5 Lamb-wave sensor systems

Several groups have studied the generation and detection of acoustic waves in zinc oxide/silicon layered structures. There are several wave types used for sensor applications but not all

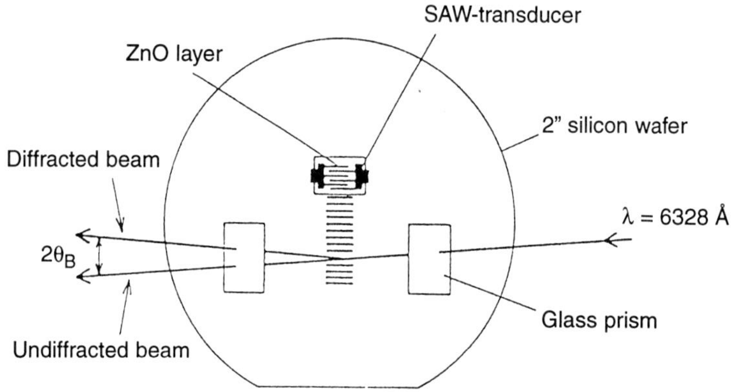

Figure 3.20. *Top view of the silicon implemented acousto-optic deflector (after Benoist et al. [3.42]).*

of them are easy to integrate in silicon. Completely integrated Rayleigh wave and Lamb-wave oscillators have been realized in ZnO-Si layered structures [3.43–3.44]. In section 3.3 the fabrication sequence of a monolithic zinc oxide covered micromachined Lamb wave sensor was described. Of course, the acoustic part and the electronic device can also be fabricated on two separate silicon chips, forming a hybrid set-up. Advantages of the monolithic setup are found in the oscillator stability (because of the short distance between acoustic device and amplifier), and the small size of the complete system. The energy dissipation of the electronic circuitry causes a rise of operating temperature of the complete chip. This can be a disadvantage in some applications.

The oscillator system has some properties which make it very attractive for use as a sensor. Firstly, the front, where the electronics are located, and the back, where the sensor contacts the fluid, are separated perfectly (Fig. 3.21). Secondly, the sensor has two outputs, one being the oscillator frequency and the other the gain voltage, which is related to the acoustic losses in the ZnO-Si layered plate. Both wave velocity and amplitude are monitored in this way; this gives independent information about different properties

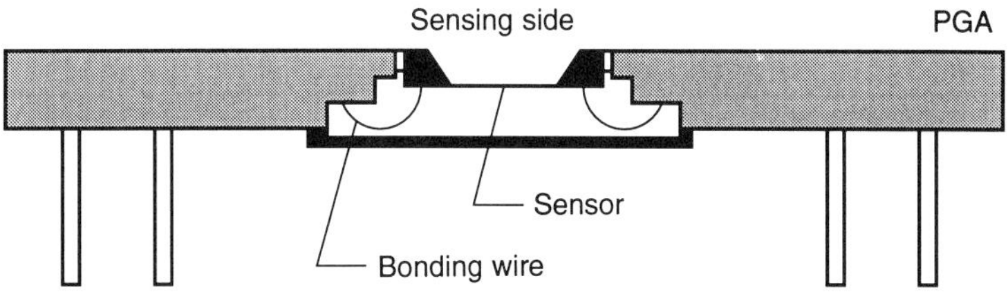

Figure 3.21. *Silicon Lamb-wave sensor system mounted in a pin-grid array (PGA).*

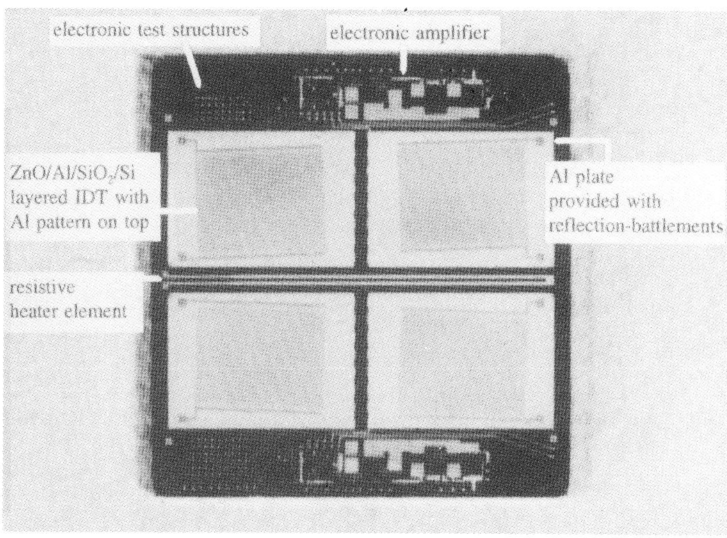

Figure 3.22. *Photograph of the monolithic Lamb-wave sensor.*

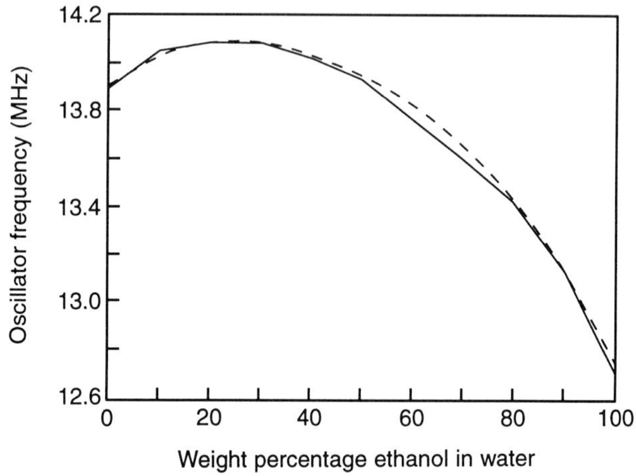

Figure 3.23. *Experimental (solid curve) and calculated (dashed curve) frequency responses to water-ethanol solutions.*

of the fluid which contacts the sensor. The sensor outputs are directly connected to a μ-controller which facilitates the use of the sensor in process-control.

In Fig. 3.22 a photograph of the monolithic Lamb-wave sensor is shown.

The system has many applications. Examples are (bio)chemical sensing by monitoring the selective adsorption of molecules at a selective chemically active thin film, the determination of liquid density and viscosity, and the determination of the composition of liquid mixtures. In Fig. 3.23 the experimental and calculated frequency responses to different water-ethanol mixtures are shown.

3.5 Discussion and conclusions

Zinc oxide is a widely investigated transducer material. It is fabricated, patterned and provided with metal electrodes by using standard planar processes. It can readily be combined with both bipolar and CMOS IC-processes, not least because the deposition and further processing can be performed after the IC-process has (almost) been completed.

At low-frequency or DC applications, the relatively high conductivity of the zinc oxide might cause current leakage. This can be prevented by depletion of the zinc oxide film or by using interjacent dielectric layers.

Piezoelectric thin films of zinc oxide are not extremely stable. The films absorb water, which alters the elastic and piezoelectric properties and damages the film (aging). Also, the internal stress in zinc oxide films which occurs during growth is a matter for concern. The stress does not only affect the performance of the transducer, but in time, relaxation of the stress will occur, again affecting the performance (e.g. reproducibility). These problems are solved by using a reproducible low-stress zinc oxide fabrication process and by careful packaging of the zinc oxide-containing devices. Also, post-deposition annealing and passivation of the zinc oxide by a thin silicon nitride film is possible [3.27].

The standardization of the zinc oxide processes in IC-facilities would further increase the number of applications of zinc oxide on silicon devices.

3.6 References

[3.1] L.J. Maissel and R. Glang, *Handbook of thin film technology*, McGraw-Hill, New York, 1970, chapter 15.

[3.2] A. Venema, Fabrication of CdS patterns on Si substrates by standard planar integrated circuit technology, *Thin Solid Films*, 22 (1974) S27–S29.

[3.3] A.J. Bahr, R.E. Lee, F.S. Hickernell, C.B. Willingham, T.M. Reeder, A comparison between the physical structure of sputtered zinc oxide films and the measured electromechanical coupling factor for surface acoustic waves generated by such films, Proc. *IEEE 1972 Ultrasonics Symp.*, 202–205.

[3.4] F.S. Hickernell, dc triode sputtered zinc oxide surface elastic wave transducers, *J. Appl. Phys.*, 44(3) (1973) pp. 1061–1071.

[3.5] B.T. Khuri-Yakub, G.S. Kino and P. Galle, Studies of the optimum conditions for growth of rf-sputtered ZnO films, *J. Appl. Phys.*, 46(8) (1975) 3266–3272.

[3.6] K.M. Lakin, J. Liu, K. Wang, Aluminum nitride on sapphire, *IEEE Proc. 1974 Ultrasonics Symp.*, pp 302–306.

[3.7] A. Fiorillo, P. Dario, J. van der Spiegel, C. Domenici, J. Foo, Spinned P(VDF-TrFE) copolymer layer for a silicon-piezoelectric integrated US sensor, *Proc. IEEE 1987 Ultrasonics Symp.*, Denver, pp. 667–670.

[3.8] R.G. Schwarz, J.D. Plummer, Integrated silicon-PVDF acoustic transducer arrays, *IEEE Trans. on Electron. Devices*, ED-26 (12) (1979) 1921–1931.

[3.9] P.M. Sarro, Sensor technology strategy in silicon, *Sensors and Actuators*, A31 (1992) 138–143.

[3.10] C. van Mullum Micromachined silicon integrated resonant sensors, PhD thesis, Twente University, 1993, chapter 5.3.

[3.11] F.R. Blom, D.J. Yntema, F.C.M. van der Pol, M. Elwenspoek, J.H.J. Fluitman and Th.J.A. Popma, Thin-film ZnO as a micromechanical actuator at low frequencies, *Sensors and Actuators*, A21–23 (1990), 226–228.

[3.12] C. van Mullum, Micromachined silicon integrated resonant sensors, PhD thesis, University of Twente, 1993, chapter 4.

[3.13] Ph. Defranould, High deposition rate sputtered ZnO thin films for BAW and SAW applications, *Proc. IEEE 1981 Ultrasonics Symp.*, pp. 483–488.

[3.14] F.S. Hickernell, Microstructure of ZnO films used for acoustic surface-wave generation, *J. Vac. Sci. Technol.*, 12(4) (1975) pp. 879–883.

[3.15] F.S. Hickernell, Surface acoustic wave propagation loss in zinc oxide films, *Proc. IEEE 1982 Ultrasonics Symp.*, pp. 325–328.

[3.16] T. Yamamoto, T. Shiosaki, A. Kawabata, Characterization of ZnO piezoelectric films prepared by rf planar-magnetron sputtering, *J. Appl. Phys.*, 51(6) (1980) 3113–3120.

[3.17] S. Sen, D.J. Leary, C.L. Bauer, Characterization of rf-sputtered ZnO thin films by X-ray diffraction and scanning electron microscopy, *Thin Solid Films*, 94 (1982) 7–14.

[3.18] O. Yamazaki, T. Mitsuyu, K. Wasa, ZnO thin-film devices, *IEEE Trans. on SU*, 27(6) (1980) 369–379.

[3.19] G.L. Dybwad, C-axis orientation of sputtered ZnO films,

J. Appl. Phys., 42(12) (1971) 5192–5194.

[3.20] J.S. Schoenwald, K.L. Keester, E.J. Staples, Characterization of rf sputtered ZnO and ZnS thin films for SAW transducers by X-ray diffraction analysis, *Proc. IEEE 1979 Ultrasonics Symp.*, pp. 926–931.

[3.21] F.R. Blom, S. Bouwstra, J.H.J. Fluitman, M. Elwenspoek, Resonant silicon beam force sensor, *Sensors and Actuators*, 17 (1989) 513–519.

[3.22] B.T. Khuri-Yakub, J.G. Smits, T. Barbee, Reactive magnetron sputtering of ZnO, *J. Appl. Phys.*, 52(7) (1981) 4772–4774.

[3.23] R.S. Wagers, G.S. Kino, P. Galle, D. Winslow, ZnO acoustic transducers utilizing crystalline gold substrata, *Proc. IEEE 1972 Ultrasonics Symp.*, pp. 194–197.

[3.24] A.D. Sathe and E.S. Kim, Techniques to control residual stress in ZnO films, *Proc.of the 7th Int. Conf. on Solid-State Sensors and Actuators, Transducers '93*, Yokohama, Japan, pp. 158–161.

[3.25] F.S. Hickernell, T.S. Hickernell, F.J. Hickernell, The etched surface of sputtered zinc oxide and its relation to fractal growth and surface wave properties, *Proc. IEEE 1992 Ultrasonics Symp.*, pp. 373–376.

[3.26] B.D. Cullity, *Elements of X-ray diffraction*, Addison-Wesley Publishing Company Inc., third printing, February 1967.

[3.27] M.J. Vellekoop, C.C.G. Visser, P.M. Sarro, A. Venema, Compatibility of zinc oxide with silicon IC processing, *Sensors and Actuators*, A21–23 (1990) 1027–1030.

[3.28] S. Chang, D.B. Hicks, R.C.O. Laugal, Patterning of zinc oxide thin films, *Techn. Digest IEEE Solid-State Sensor and Actuator Workshop*, June 1992, Hilton Head Island, pp. 41–45.

[3.29] B.A. Movchan, A.V. Demchishin, Study of the structure and properties of thick vacuum condensates of nickel, titanium, tungsten, aluminum oxide and zirconium dioxide, *Phys. Met. Metallogr.*, 28(4) (1969) 83–90.

[3.30] J.A. Thornton, Influence of apparatus geometry and deposition conditions on the structure and topography of thick sputtered coatings, *J. Vac. Sci. Technol.*, 11 (1974) 666–

670.

[3.31] F.S. Hickernell, T.S. Hickernell, Stress relaxation in triode sputtered zinc oxide films, Proc. *IEEE 1983 Ultrasonics Symp.*, pp. 311–315.

[3.32] C. Rossington, A.G. Evans, D.B. Marshall, B.T. Khuri-Yakub, Measurements of adherence of residually stressed thin films by indentation. II. Experiments with ZnO/Si, *J.Appl.Phys.*, 56(10) (1984) 2639–2644.

[3.33] J.C. Zesch, B. Hadimioglu, B.T. Khuri-Yakub, M. Lim, R. Lujan, J. Ho, S. Akamine, D. Steinmetz, C.F. Quate, E.G. Rawson, Deposition of highly oriented low-stress ZnO films, *Proc. IEEE 1991 Ultrasonics Symp.*, pp. 445–448.

[3.34] F.J. Hickernell, The effect of the structure and composition of zinc oxide films on their elastic properties, *Proc. IEEE 1977 Ultrasonics Symp.*, pp. 309–312.

[3.35] R.J. Lad, P.D. Funkenbusch, C.R. Aita, Postdeposition annealing behavior of rf sputtered ZnO films, *J.Vac.Sci.Technol.*, 17(4) Jul./Aug. (1980) 808–811.

[3.36] F.S. Hickernell, Post-deposition annealing of zinc oxide films, *Proc. IEEE 1981 Ultrasonics Symp.*, pp. 489–492.

[3.37] C.J. Gawlak, C.R. Aita, Stress relief of basal orientation zinc oxide films by isothermal annealing, *J.Vac.Sci.Technol.*, A1(2) Apr./June (1983) 415–418.

[3.38] H.H.P.Th. Bekman, K.W. Benoist, J.L. Joppe, Post-deposition annealing of RF-sputtered zinc-oxide films, *Appl. Surf. Sci.*, 70/71 (1993) 347–350.

[3.39] Y. Miyasaka, S. Hoshini, S. Takahashi, Temperature compensated ZnO/SiO_2/Si thin film resonator for VHF voltage controlled oscillator, *Proc. IEEE 1985 Ultrasonics Symp.*, San Francisco, pp. 346–351.

[3.40] D.L. Polla, W.T. Chang, R.S. Muller and R.M. White, Integrated zinc oxide on silicon tactile-sensor array, *Proc. Int. Electron Device Meeting (IEDM '85)*, Los Angeles, 1985, pp. 133–136.

[3.41] L. Kuhn, M.L. Dakss, P.F. Heidrich and B.A. Scott, Deflection of an optical guided wave by a surface acoustic wave, *Appl. Phys. Lett.*, 17 (1970) 265–267.

[3.42] K.W. Benoist, J.L. Joppe, M.J. Vellekoop and M.K. Smit, Acousto-optic light deflection in an Al_2O_3 optical waveguide structure, *Fiber and Integrated optics*, 8 (1989) 249–253.

[3.43] M.J. Vellekoop, E. Nieuwkoop, J.C. Haartsen and A. Venema, A monolithic SAW physical-electronic system for sensors, *Proc. IEEE 1987 Ultrasonics Symp.*, Denver, pp. 641–644.

[3.44] M.J. Vellekoop, G.W. Lubking, P.M. Sarro and A. Venema, Integrated-circuit-compatible design and technology of acoustic-wave based microsensors, *Sensors and Actuators*, A44 (1994) 249–263.

4
ECONOMICS OF ON-CHIP FUNCTIONAL INTEGRATION

4.1 Introduction

So far the merits and limitations of functional integration of silicon sensor and microelectronic circuits in silicon have been mainly presented from a technological perspective. At this stage the economics of the device need to be discussed. The obvious question to be asked when reviewing the wide range of integrated silicon smart sensor prototypes already fabricated at universities and research institutes is about their commercialization. For various reasons, the transition from academic research to industrial application is rarely completed, despite the many organizations established by national governments and international institutions to facilitate technology transfer from university to industry. Three issues are crucial for answering this question and for understanding the complications encountered in the commercialization of on-chip integrated silicon sensors. The first is the market; costs for entering the market are high and the size and diversity generally does not provide much incentive for mass production. The second is the device; it will be shown that, unlike microelectronic circuits, costs for wafer processing and their yield are relatively unimportant when also considering the costs of testing, bonding and non-standard packaging of a relatively small series. The third is the sensor industry in relation to the microelectronic circuit manufacturing industry.

The first two issues are important as arguments to resolve the widespread misconception that wafer yield would be the cost-determining factor. As will be shown, wafer yield is indeed the dominating cost in microelectronic circuit fabrication; however, the extensive testing of electrical and non-electrical parameters, the limited volume due to the fragmented market and the non-standard packaging of sensors cause other cost factors to be dominant. This improper use of the wafer yield criterion has too often wrongly caused the integrated silicon smart sensor to be considered as a marginal product.

The first and third issues have a direct impact on the partition of the various functions in a data-acquisition system. The on-chip integrated silicon smart sensor aims towards a maximization of on-chip functions. Obviously, technical constraints sometimes cause on-chip integration of all functions to be impossible, viz. in the case where the devices are fundamentally non-fabrication compatible. Sometimes the application imposes constraints that limit the optimal level of functional integration, such as in biomedical applications, where biocompatibility is impaired by heat dissipation, as will be outlined in chapter 6. However, it will also be shown that economic considerations strongly affect the partitioning that is considered optimum. At present the silicon sensor industry is composed of medium-sized companies that do not have in-house standard microelectronic circuit production. As mentioned in the previous chapters, the future of silicon sensors depends largely on the extent to which the advantages of the material compatibility with microelectronic circuits are actually to be exploited. In the extreme, this implies on-chip integration of the complete data-acquisition system.

Such a level of integration requires a processing control that is commonplace in state-of-the-art integrated silicon circuit manufacturing, but not in the silicon sensor industry. Moreover, a more-universally applicable data-acquisition unit would result in a market volume that would attract the interest of the microelectronics industry. Thus, the trend towards increased on-chip integration does not benefit the sensor industry, which, therefore, favors a lower level of integration. Such an attitude delays the introduction of on-chip functional integration and might harm the longer-term competitiveness of the sensor industry. Therefore, the relation between the sensor industry and the microelectronics manufacturers needs to be investigated. These three issues will be discussed in more detail in the next sections.

4.2 The economies of scale and the wafer yield

The unit costs of a product are determined by the investment that is required to start production divided by the expected number of units that are to be fabricated before the investment is fully depreciated (fixed costs), plus the costs for the production of each individual unit (variable costs). Figure 4.1 shows three typical curves for the unit costs at an increased degree of investment [4.1]. In the case of hand-made sensors, no investment is made and the costs per unit remain the same irrespective of the number of units already fabricated (line a). Of course this is a highly simplified situation as some investment is always required for tools, buildings, etc. and a craftsman is likely to gain experience that would enable him/her to increase productivity in time. At an increasing level of investment (curve b), labor costs for each unit decrease relative to a hand-made product, but fixed costs are added depending on production volume. Therefore, the costs per unit decrease with production volume until, at extremely high production volumes, only the (reduced) variable costs apply. From these curves it can be concluded that, apart from quality control, mass production and labor costs are the main incentives for industrial investment. In the semiconductor manufacturing industry, this mechanism is pushed to the extreme, as indicated in curve (c). Microelectronic circuit production is only economically viable for a very large production volume ($>10^6$ units). Such a volume is commonplace in integrated circuits, but is rarely achieved in the sensor industry. The reasons for the limited production volume in sensors will be discussed in more detail in section 4.3.

A microelectronic foundry is designed to enable the processing of a fixed number of wafers during its economic lifetime. Although the wearing-out of machines is a consideration in the number of wafers for which the foundry is specified, the capacity, in terms of number of wafers that can be fully processed per unit time, in

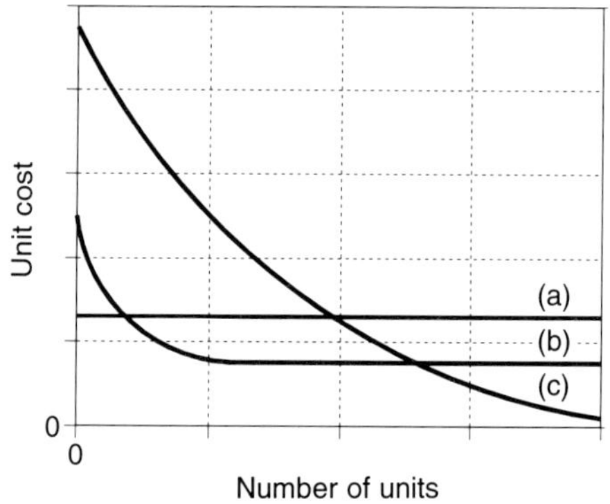

Figure 4.1. *Unit costs vs. production volume at an increasing investment.*

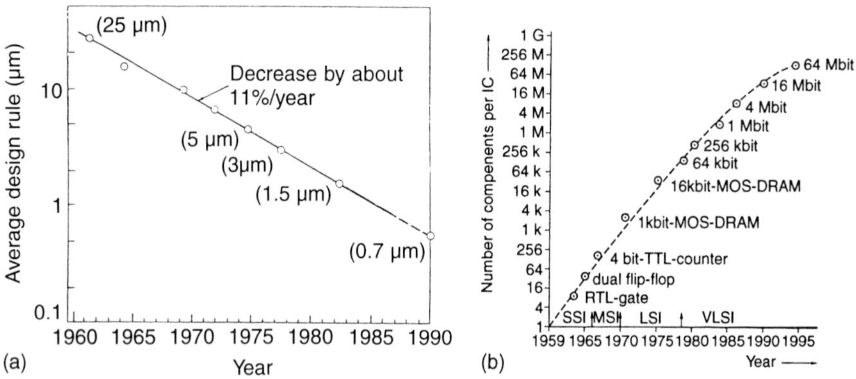

Figure 4.2. *Innovation rate in the microelectronics industry: (a) minimum feature size and (b) number of active components vs. time.*

combination with the very limited time-span that such a foundry can be considered as the most-advanced available, due to the high innovation rate in the microelectronics industry, is usually more restrictive. The innovation rate is usually estimated from the reduction of the minimum feature size and the increasing number of integrated components in time, as shown in Fig. 4.2 [4.2–4.3].

If the number of wafers that can be processed during the lifetime of a foundry is fixed, then the economics of the entire operation

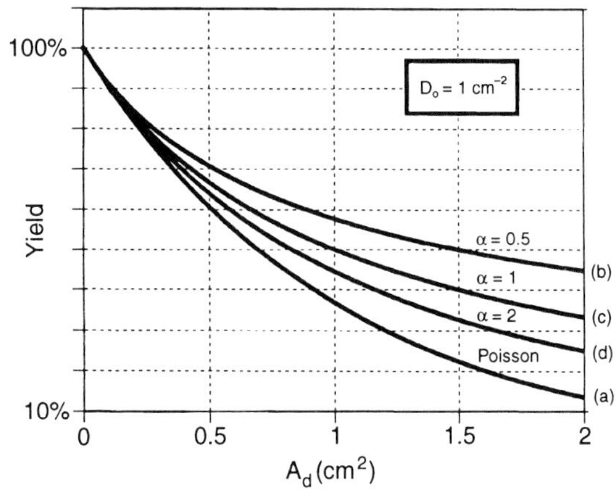

Figure 4.3. *Wafer yield vs. die area for* $D_o = 1\ cm^{-2}$ *using (a) the Poisson distribution or the gamma function with (b)* $\alpha = 0.5$, *(c)* $\alpha = 1$ *and (d)* $\alpha = 2$.

is set by (a) the size of the investment and (b) the number of working devices that can be realized out of a wafer; the wafer yield. Consideration (a) determines whether the investment could pay-off and will become an issue for the next generation of silicon foundries for each major microelectronic circuit manufacturer, whereas (b) sets the price and is a figure of merit for the design, the technology used and the process control enforced, and determines the competitiveness of each individual manufacturer.

Corporate investment policy for the next generation of foundries will include decisions on global cooperation, specialization and joint ventures and is not discussed here. The second factor, the wafer yield, Y, is also relevant for integrated sensors and is defined as the probability of zero defects over the die area, A_d. When assuming a random defect distribution over the wafer, the Poisson yield model applies: $Y = \exp(-D_o A_d)$, where D_o denotes the average defect density defined as the number of defects per wafer, N_d, over the wafer area, A_w: $D_o = N_d/A_w$. As defects are not randomly distributed over the wafer, but rather clustered, the Poisson yield model usually provides too pessimistic a prediction. More realistic yield predictions are made using the gamma function instead, with a clustering parameter α, giving the expression:

$$Y(A_{\mathrm{d}}) = \left(1 + \frac{D_{\mathrm{o}}A_{\mathrm{d}}}{\alpha}\right)^{-\alpha} \tag{4.1}$$

Note that for $\alpha \rightarrow \infty$ the expression gives the Poisson yield model [4.3]. Practical processes result in $0.5 < \alpha < 4$, where the smaller values of α indicate a higher degree of process maturity [4.4][4.5]. In Fig. 4.3 the yield is given as a function of $D_{\mathrm{o}}A_{\mathrm{d}}$ for various models. The yield drops with die area and generally $D_{\mathrm{o}}A_{\mathrm{d}} \ll 1$ for an economically viable process.

The validity of wafer yield modeling in microelectronic circuit manufacturing has also resulted in its implementation for estimating the economic viability of the smart sensor concept. This adoption of the wafer yield criterion from the microelectronics industry can be explained by their common technological roots.

4.3 The sensor market

The rapid market acceptance of digital and analog integrated circuits was largely due to the high level of abstraction of the internal operation. Such a feature is also essential for the market penetration of the silicon sensor. The silicon sensor should either have a superior specification compared to e.g. a *de facto* industrial standard such as the Pt100 temperature sensor, or the inherent advantages of the material compatibility with integrated circuits in silicon should actually be exploited. Fabrication compatibility is required to convert the basic silicon sensing element into an integrated data-acquisition subsystem (integrated silicon smart sensor), which supplies a standard digital output signal, not corrupted by sensor non-idealities, over a sensor bus to a system controler. The internal functional complexity of the integrated silicon smart sensor should be sufficient to (a) allow implementation in a system without knowledge about the details of the internal operation, and (b) minimize overhead to the system controler so that the sensor can be considered as a complete data-acquisition subsystem in a measurement and control system.

However, the integrated silicon smart sensor will be a component of a measurement and control system. The specifications of such a system are usually strongly application-specific, which has resulted in a fragmented market. Therefore, unlike the market for consumer electronics in which many microelectronic circuits are used, the instrumentation and control market is a low-volume market.

Moreover, sensor packaging lacks standardization, due to the diversity of the market and the application-dependence of the package. Different requirements are, for example, imposed on a pressure sensor package in a biomedical application (measuring the pressure in the coronary artery) and in automotive applications [4.6]. The pressure-sensing elements could well operate in the same pressure range and originate from the same design. Adding the volume of all applications together would lead to the conclusion that pressure sensing is one of the few sensor areas where mass production on the wafer level is achieved. The present annual fabrication volume in Europe amounts to 14 million and forecasts predict 35 million in 5 years [4.7]. However, the wide range of different packages makes the fabrication of packaged and tested pressure-sensing devices still a small-volume operation.

This structure of the sensor market has strong implications for the economic viability of the smart sensor concept. On the one hand the smart sensor concept provides a means to realize a more-universal building block on-chip with an inherently larger range of applications. On the other hand, each application imposes different demands that cause the packaged sensor to be different in each application.

4.4 Wafer yield economics in on-chip integrated sensors

Yield modeling can be used to explain the most popular prejudice against the silicon smart sensor concept. It is generally assumed that the merging of sensor and readout circuits increases the costs due to the increased total die area, similar to the decreasing yield and the resulting increasing unit costs that occur for an integrated circuit with an increasing die area. Obviously, increasing the active die area results in a reduced yield irrespective of the model, so the Poisson model is used. Adding a sensor of die area A_s to an integrated circuit consuming a die area A_c, thus giving a total die area $A_s + A_c$, leads to a device with a poorer fabrication yield; $Y_{ss} = \exp[-D_o(A_s + A_c)] < \exp(-D_o A_c)$. Adding a sensor area on-chip would also reduce the yield. However, the readout electronics has to be included one way or another, so the yield of an integrated smart sensor should more properly be compared with that of a hybrid device, Y_{hs}.

In a hybrid smart sensor, a sensor chip of die area A_s is processed in a process with defect density D_{os} and is put in a package with a chip containing the microelectronic circuits of die area A_c and defect density D_{oc}. Electrical connection between the two chips is made using wire bonding. Assuming the wire bonding does not introduce a yield loss, the defect densities in the sensor wafer and the circuit wafer are statistically independent and applying the Poisson yield model gives: $Y_{ss} = \exp[-D_o(A_s + A_c)]$ and $Y_{hs} = \exp(-D_{os}A_s) \cdot \exp(-D_{oc}A_c) = \exp(D_{os}A_s + D_{oc}A_c)$. Obviously, $D_{os}A_s < 1$ and $D_{oc}A_c < 1$ in a commercial process. An integrated silicon smart sensor is only economically viable if $Y_{ss} > Y_{hs}$. However, merging a sensor process with a microelectronic process usually results in an increased complexity and thus in an increasing defect density: $D_o \geq \max(D_{os}, D_{oc})$. Therefore, $D_o(A_s + A_c) > D_{os}A_s + D_{oc}A_c$ and this general opinion would indeed be confirmed for any combination of A_s and A_c.

However, as mentioned already, defects are clustered and the Poisson yield model has already been abandoned in the microelectronics industry. Assuming a reasonably mature process ($\alpha \simeq 1$) results in, for the integrated silicon smart sensor in: $Y_{ss} = [1+D_o(A_s+A_c)]^{-1}$ and for the hybrid sensor: $Y_{hs} = [(1+D_{os}A_s)(1+D_{oc}A_c)]^{-1}$. Again, merging two processes of a certain defect density results in a compatible process with a defect density larger than in any of the basic processes: $D_o \geq \max(D_{os}, D_{oc})$. Various types of smart sensors can now be analyzed with respect to the economic viability based on wafer processing yield.

The first case is the integrated thermal sensor where $D_o = D_{os} = D_{oc}$ and $A_s < A_c$. It only makes sense to fabricate an integrated silicon smart sensor when $Y_{ss} > Y_{hs}$, so $\alpha = 1$ results in: $D_o < D_{os}A_s/(A_s + A_c) + D_{oc}A_c/(A_s + A_c) + D_{os}D_{oc}A_sA_c/(A_s + A_c) = D_{oc}[1+D_{oc}A_s/(1+D_{oc}A_s)]$. Again, technological constraints impose $D_o \geq \max(D_{os}, D_{oc}) = D_{os} = D_{oc}$. Both restrictions are, therefore, met in an integrated thermal smart sensor and a commercial integrated silicon thermal smart sensor based on the design depicted earlier in Fig. 1.21 is technologically and economically feasible when considering only wafer processing.

The second type of integrated silicon smart sensor of special interest is the simple smart sensor, such as the thermopile-based flow sensor shown earlier in Fig. 1.22, which is characterized by $D_{oc} < D_{os}$ and $A_s > A_c$, thus $Y_{ss} > Y_{hs}$ for $D_o < D_{os}A_s/(A_s + A_c) + D_{oc}A_c/(A_s + A_c) + D_{os}D_{oc}A_sA_c/(A_s + A_c) \simeq [D_{os}A_s/(A_s + A_c)](1 + D_{oc}A_c) < D_{os}$. This condition is not in agreement with $D_o \geq \max(D_{os}, D_{oc})$, so a simple smart sensor is not economically viable according to the yield criterion.

Alternatively, a complex integrated silicon smart sensor is characterized by $D_{oc} < D_{os}$ and $A_s < A_c$, thus $Y_{ss} > Y_{hs}$ for $D_o < D_{os}A_s/(A_s + A_c) + D_{oc}A_c/(A_s + A_c) + D_{os}D_{oc}A_sA_c/(A_s + A_c) \simeq D_{os}A_s/(A_s + A_c)(2 + D_{os}A_s)$. Both the technological and economic constraints can be satisfied for a limited range of A_c, A_s, D_{oc}, D_{os}. Generally, from the yield viewpoint it can be concluded that when one takes the trouble to merge sensor and readout electronics it is better not to restrict to a few transistors, but rather to make it worthwhile. Figure 4.4 indicates the requirements for economic integrated silicon smart sensor fabrication for $\alpha = 1$ and typical values of D_{oc}/D_{os}.

The problem in applying the yield models developed in the microelectronics industry is the fact that these only lead to a sensible economic criterion in mass production. As indicated in the previous section, this is rarely achieved in sensors. Furthermore, it has to be emphasized that the cost of any microelectronic device is composed of wafer processing, wafer level testing, dicing, bonding, packaging and final component testing. In microelectronic circuits almost all of the testing can be done on the wafer level and only little yield loss is detected at the final component testing. Automated wafer testing is very efficient and ensures a high throughput with little unnecessary further processing of defective devices. Moreover, non-electrical testing is merely required to verify the environmental operating range specification, such as temperature and humidity, so costs of testing are relatively low. Relatively inexpensive standardized molded plastic packages are

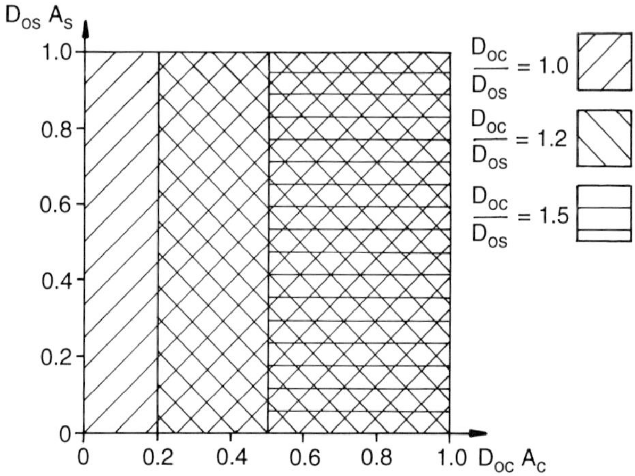

Figure 4.4. *Comparing the smart sensor with the hybrid realization in a process with* $\alpha = 1$, $Y_{ss}/Y_{hs} \geq 1$ *in the hatched areas.*

used for encapsulation. Wafer processing contributes, therefore, to a significant part of the fabrication costs in microelectronic circuits.

For integrated silicon smart sensors, the contribution of the different types of costs is quite different. Extensive testing is required on the packaged device for two reasons. Firstly, as a sensor is converting signals from a non-electrical domain into electrical signals, unlike microelectronic circuits, both electrical and non-electrical testing are required. The packaging often influences the performance, such as encapsulation-stress-induced offset in a Hall plate, or the package is an essential part of the sensor structure, such as the cap on the reference cavity in an absolute pressure sensor or the capped wafers for determining the damping in an accelerometer. Therefore, most of the testing has to be done on packaged devices and adds a much larger share to the fabrication costs in an integrated smart sensor compared to an integrated microelectronic circuit.

Packaging is a special problem in sensors. The application dependence of the package and its economic implications have already been discussed in the previous section. Microelectronic circuits are routinely packaged in standard molded plastic in order to provide sufficient protection against environmental influences, such as moisture and mechanical stress. Integrated circuit packaging is, therefore, a high-volume operation. Moreover, the integrated circuit package has been designed solely to provide an effective isolation from external hazards; however, a sensor requires by definition some degree of environmental access. As mentioned already, thermal sensors are the least demanding with respect to packaging and the problems are the most severe in chemical sensors.

Evaluation of the different types of costs in the fabrication of integrated silicon smart sensors leads to the conclusion that its merit can hardly be discussed in terms of wafer processing yield. The popular misconception, that the yield with two separate dies must always be better than the yield of the merged device, which consumes the same total area, no longer holds.

Also the limited production volume in integrated sensor fabrication and the fact that wafer processing costs amount only to a fraction of the total costs, due to the larger share of testing and packaging, make the wafer yield approach and the emphasis on die area minimization an inadequate means for estimating the economic viability of an integrated silicon smart sensor. An analysis in terms of added value would be more appropriate.

4.5 The 'added value' concept

Any commercial operation on a product, whether manufacturing, storage or transportation, adds to its value. Generally, storage and transportation (transaction costs) are included in between each manufacturing step, so for simplicity the discussion is limited to the clearly distinguishable manufacturing steps. Usually, the market sets the price of the complete system, such as the Anti-lock Braking System (ABS) or the airbag. The price is usually based on what is considered a reasonable fraction of the total cost of the particular consumer product (the automobile) and is in the order of \$100 at the time of introduction of such a system on the market. This market mechanism is superseded in the case of government regulation on safety or exhaust emission. Tracing back the conventional manufacturing procedure (component fabrication, subsystem assembly and system integration) results in a maximum allowable cost for basic sensing element fabrication in the $1-range. This price tag is often also wrongly attached to integrated silicon smart sensors, which are basically complete data-acquisition subsystems and are, therefore, on a higher hierarchical system level.

Obviously, an integrated silicon smart sensor should not have to compete with a general purpose integrated circuit. Many integrated circuits have to be used in combination with an elementary sensing element to realize a conventional data-acquisition subsystem on the PC-board level. The latter device is a system of higher added value and the actual competitor of the integrated silicon smart sensor, whose economic viability should be determined from a comparison with the costs of packaged elementary sensors and external readout and selection circuits, including wire bonding and assembly on a PC board.

Not every integrated silicon smart sensor that is technically feasible will meet the economic constraints for commercial success (even if this were solely determined by objective criteria such as costs). The magnitude of the added value is an application-specific uncertainty, as reference has to be made to the state-of-the-art in that specific field based on conventionally fabricated devices. Generally, the added value is maximized by converting the sensing element into a data-acquisition subsystem that provides a bus-organized digital output code of standard format free from any sensor non-idealities. The complexity of the bus interface should be sufficient to allow the system controler to regard the data-acquisition unit as a data source of a high-level of abstraction that does not cause much overhead and is equipped with self-test, self-calibration and an error handling capability. These requirements, combined with the limited impact of wafer processing costs on the total costs of integrated silicon smart sensors, makes it undesirable to compromize on complexity solely for sake of the wafer yield.

4.6 The economics of special compatible post-processing

The added value criterion is of particular interest when trying to estimate the economic viability of special compatible sensor technologies in silicon or the deposition of compatible materials

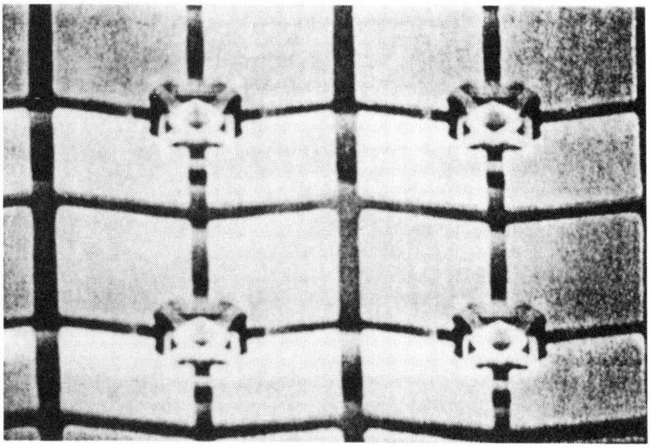

Figure 4.5. *Electrostatically deflectable mirror array.*

on a silicon substrate. Such technologies have to be resorted to when simple modifications in a standard process are not sufficient to achieve proper silicon sensor operation. Micromachining is such a technology and has been outlined in chapter 2. ZnO has been introduced as a compatible material in chapter 3. In this section the economics of such special compatible processes or materials will be discussed and it will be shown that the viability depends solely on the amount of added value that results from the application.

Bulk micromachining of the silicon underneath a thermopile has been shown to be essential for proper operation of Seebeck effect-based thermal sensors. The added value of the micromachining step is, therefore, significant. Either micromachining should be applied or silicon cannot be seriously considered for this application.

A more subtle example involves the economics of merging surface micromachining for fabrication of an array of mechanical sensors or actuators and on-chip circuits. Surface micromachining has been employed for the fabrication of large arrays of torsional mirrors. These are composed of polysilicon plates covered by reflective metal and suspended in the center by a beam as shown in Fig. 4.5 and are intended for application in image projection [4.8]. Applying a voltage between either side of the plate and the substrate causes electrostatic tilting of the plate towards the side where the voltage is applied and thus the projection of incident light can be controled to result in a 2-dimensional image on a screen. This application requires integration in a smart silicon actuator, owing to the need for on-chip multiplexing in order to limit the number of interconnects, while enabling the addressing of each individual deflection element. The added value of the smart sensor concept is, therefore, considerable in this application.

The third example concerns Selective Epitaxial Growth (SEG) and Epitaxial Lateral Overgrowth (ELO), which offer opportunities to alleviate compatibility problems when a vertical doping profile is required for the sensing element that is completely different from the profile that results from the standard microelectronic process.

SEG takes place on parts of a silicon wafer that are not covered by a dielectric layer (nitride or thermally grown or deposited oxide) [4.9]. These seed holes are usually defined by the contact holes. Under special conditions, $T = 950°C$, $P = 100$ Torr and using the SiH_2Cl_2–HCl–H_2 gas system, crystaline growth takes place through the seed windows and proceeds over the dielectric, after reaching the surface, to form an Epitaxial Lateral Overgrown (ELO) silicon layer. Figure 4.6 shows the SEG and ELO structures.

SEG of silicon basically enables the stacking of junctions in addition to those resulting from the standard process.

A vertical doping profile can be realized in the ELO layer that is optimized for the sensing element irrespective of the active devices integrated already in the substrate. A 2 μm thick SEG layer requires heating to 950°C for 30 min. Although this addition to the thermal budget cannot simply be disregarded, the effect on active bipolar and CMOS components is usually acceptable, as has already been discussed in chapter 1.

Crystal quality of the SEG is dependent on seed hole size and orientation. Diodes in SEG indicate an almost ideal performance of the (100)-oriented device (ideality factor is 1.08). Layer thickness uniformity over the wafer is limited by masking effects to 6% [4.10]. Defects are mainly formed along the SEG layer circumference. Research on SEG has so far mainly been directed at sub-micron high-speed active devices. Due to the high defect density resulting from the relatively large circumference/area ratio, SEG technology has never been very successful in this application. However, this is

(a)
(b)

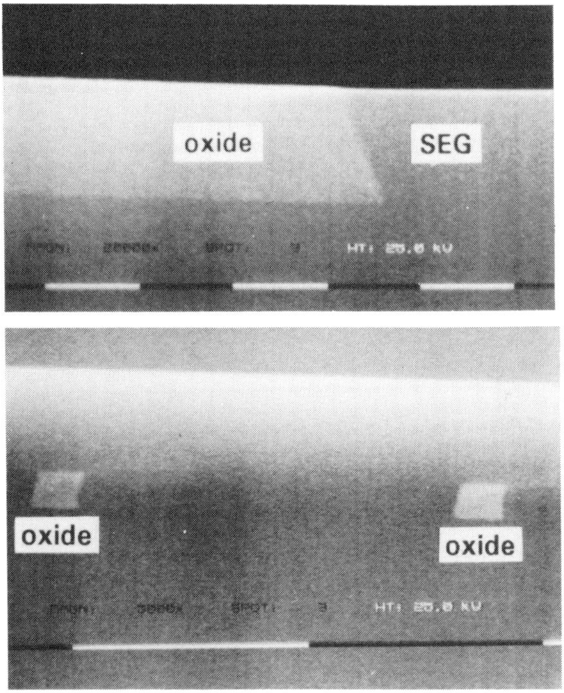

Figure 4.6. *(a) (100) SEG layer in a contact window. (b) ELO layer grown from two contact window strips spaced 2 μm apart.*

due to the very small seed holes mandatory in those devices. SEG is expected to become a successful technology in integrated silicon smart sensors.

ELO layers of high crystaline quality are, in principle, formed by unrestrained growth of the silicon over the dielectric surface and a subsequent planarization. Surprisingly few defects are formed at the interface when the overgrown layers meet due to the growth at 950°C. As the vertical and lateral growth rates are about equal, improved-quality ELO layers are fabricated using confined growth in a surface-micromachined polysilicon cavity [4.9]. ELO enables the electrical and structural separation between differently doped parts—the substrate containing the bipolar readout circuits and the overgrown layer the sensing element, or *vice versa*—while main-

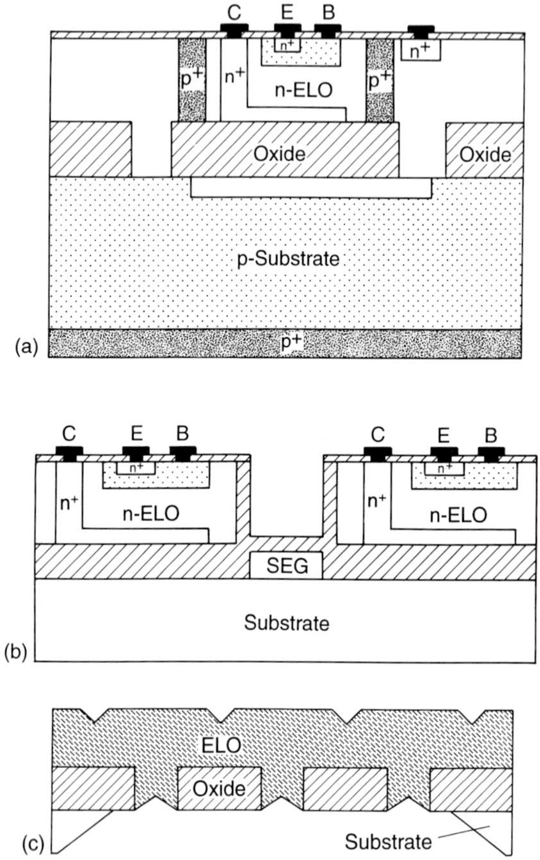

Figure 4.7. *Applying ELO in (a) readout of high-energy radiation detector array, (b) SOI and (c) bulk micromachining.*

taining the physical and functional unity of the integrated silicon smart sensor. This principle is being investigated for application in an integrated smart high-energy radiation detector, as shown in Fig. 4.7a, where the bipolar devices are fabricated in the overgrown layer on a high-purity substrate. ELO is also investigated for application in Silicon On Insulator (SOI). Plasma etching of the original seed windows is performed to ensure complete isolation from the substrate, as shown in Fig. 4.7b. Finally, the buried oxide layer can be used as an etch-stop in bulk micromachining to result in devices as depicted in Fig. 4.7c [4.11].

At this stage it is most interesting to evaluate the added value of the three SEG/ELO applications mentioned. The bonding problem is excessive in radiation detectors, where typically up to 1000 detectors with 10-100 μm pitch are integrated on a high-ohmic substrate. The high resistivity of the wafers seriously complicates the integration of amplifiers and multiplexers in the same substrate. The added value of ELO is, therefore, extremely high and the merging of standard processing with ELO is worthwhile to pursue, despite the technological difficulty of ELO.

SOI devices on a silicon substrate have already been demonstrated using laser recrystalization, porous silicon formation and oxidation and high-dose oxygen implant with a subsequent high-temperature anneal (SIMOX) [4.12]. None of these techniques is fabrication compatible with standard processing. In particular the harsh temperature treatment in SIMOX is not compatible with proper operation of already integrated devices, which would favor ELO. SIO is in fact a universally applicable technology and not tied to a particular application. The added value in general is, therefore, impossible to estimate.

The third example basically involves a novel bulk micromachining etchstop technique. Similar structures can be realized using simpler technologies, so that the added value of ELO is very small in this application.

4.7 The sensor industry

As mentioned already in the Introduction, the silicon sensor industry would, at first sight, have little reason to embrace the integrated silicon smart sensor concept. The ultimate goal of this concept is the on-chip integration of all the functions in a data-acquisition system (sensor, analog readout, ADC and digital circuits).

The integrated silicon smart sensor aims for a universal data-acquisition unit for a particular non-electrical quantity, which should also result in a production volume that attracts the interest of the microelectronics industry, whereas the level of process control required is commonplace in the microelectronics industry. Obviously, there is little incentive for the sensor industry to take the role of 'water carrier' by developing the basic sensor towards an integrated silicon smart sensor on behalf of the microelectronics industry. This rather gloomy prospect will be shown to be not quite correct.

Integrated silicon smart sensors can be classified according to their fabrication compatibility into three categories, viz. (a) those that do not impose any compatibility infringement and come automatically with a standard process (such as thermal sensors), (b) those that require special pre- or post-processing (such as micromechanical sensors), and (c) those that give rise to significant departures from standard processing, but have a huge added value. The impact of sensor manufacturing on the microelectronic process and the opportunities for the sensor and microelectronics industries should be evaluated for each category.

Category (a) of integrated silicon smart sensors is a lost cause for the sensor industry. If a mass-volume application comes up, the microelectronics industry is bound to start fabrication. There would be a future for sensor companies specialized in testing and calibration to establish the non-electrical specifications of the device.

The circumstances are more favorable for the sensor industry in the case where a device requires special pre- or post-processing to be merged with a standard microelectronic process. When considering the microelectronics and sensor industries as competitors, then sensor manufacturers initially have the advantage due to their expertise in the multi-disciplinary problems associated with the design, fabrication, testing and packaging of sensors. However, as the sensor industry usually does not have a standard production line available, an increased level of on-chip functional integration would make it increasingly more interesting for the microelectronics industry to invest in the knowledge required for sensor manufacturing. The threshold beyond which such an investment would pay off depends on many device- and application-dependent parameters, but such an investment is very limited compared to that required by a sensor manufacturer to upgrade the production facility. Basically the production volume at which the unit costs of the smart sensor crosses that of the hybrid sensor is larger for the non-IC manufacturer. This effect will be discussed in more detail for micromechanical sensors in the next chapter.

Therefore, a more rational approach for the sensor industry is to try and find a niche and use the microelectronics industry as a subcontractor to supply the microelectronically processed wafers that are subsequently to be subjected to in-house sensor processing. This approach is a direct extension of the way the infrastructure established by the microelectronics industry has traditionally been exploited by the sensor industry. The availability of this infrastructure has already been identified as one of the most important advantages of silicon sensors in terms of substrate purity and equipment.

In the case of severe compatibility infringements (category (c)), the situation is even more favorable for the sensor industry, as the microelectronics industry is usually very reluctant to change standard processes and consequently the threshold for entering such a market would be higher. This (rather small) segment of the market is, therefore, in practice exclusively for the silicon sensor industry. It is interesting to note that the economic viability of this type of device depends solely on the added value that results from the application, which causes the relatively small size of the modal sensor foundry to be an advantage. Three strategies are available to penetrate this market: (c1) have microelectronic circuits processed at a major microelectronic foundry based on your own design and subject the wafers to special sensor processing similar to devices in category b, (c2) have the circuits processed in a major microelectronic foundry, realize the sensor structure on another wafer, and merge the two in a last processing step using wafer-to-wafer bonding or flip-chip bonding, or (c3) upgrade the in-house production facility for non-standard microelectronic processing in which sensor and circuits are merged on the wafer level.

The first approach is only practical if the compatibility infringement is still manageable and can be resolved by modifying the design rules. Increasing the emitter-collector mask separation of the lateral pnp has already been discussed in chapter 1 as a method to maintain an acceptable breakdown voltage at a decreasing epilayer doping concentration. The second approach is important when the non-compatibility exceeds the compliance of the standard process. The microelectronics industry can, nevertheless, also be subcontracted in this approach by postponing the compatibility issue to the very last processing step; the bonding. The sensor wafer can be processed in-house or a sensor foundry can be subcontracted to process the sensor wafer. The wafers are bonded and the stack can be subjected to further (compatible) sensor processing. The additional advantage of this approach is that a manufacturer can choose the most suitable circuit and sensor manufacturer and perform the bonding in-house, with a minimum investment in fabrication equipment.

As already mentioned in chapter 2, care should be taken that the bonding itself does not introduce a compatibility infringement, which implies low-temperature wafer-to-wafer bonding.

Basically, wafer-to-wafer bonding enables a partitioning of the functions of a data-acquisition unit over two dies and thus does not comply with the goals of on-chip functional integration. However, this technique can also be regarded as an intermediate step towards functional integration, as the wafers are bonded to a physical unity. Moreover, its potential is the highest in the case of a significant fabrication compatibility problem and it allows the cooperation of a sensor manufacturer, who has established a niche, with a major microelectronics manufacturer, without the former having to be concerned about losing proprietary information.

The third approach is relevant in the case of a severe compatibility infringement. The sensor foundry would be able to build up a facility for the fabrication of the sensor and the non-standard readout circuits. Usually, this approach is only viable in applications where a huge amount of expert knowledge is required. This circumstance both protects the sensor manufacturer, as the microelectronics industry is very unlikely to enter such a field, but also makes such a manufacture vulnerable, as such a market does not allow for diversification.

In general it can be concluded that a sensor foundry has nothing to gain from delaying the introduction of the integrated silicon smart sensor concept. Moreover, the silicon sensor industry would improve competitiveness by introducing specialization and cooperation with microelectronic circuit industry.

4.8 Conclusions

The market, the device and the sensor industry have been analyzed in order to evaluate the economic viability of the smart sensor concept. It has been shown that the economies of scale do not apply due to the fragmented market and the variable costs caused by bonding, testing and packaging of the device. The application-dependent added value of the smart concept has been found to be a suitable criterion that can also be used to judge new sensor technologies. A suitable strategy for the sensor industry is to find a niche and to use the microelectronics industry as a subcontractor to deliver processed microelectronic wafers that are subsequently submitted to in-house sensor processing.

4.9 References

[4.1] S. Middelhoek, Quo vadis silicon sensors, *Sensors and Actuators*, A41-42 (1994) 1–8.

[4.2] S. Wolf and R.N. Tauber, *Silicon processing for the VLSI era*, Vol. 1: Process technology, Lattice Press, CA, USA.

[4.3] H.J.M. Veendrick, *From Basics to ASICs*, VCH, Weinheim, FRG, 1992.

[4.4] T.L. Michalka, R.C. Varshney and J.D. Meindl, A discussion of yield modeling with defect clustering, circuit repair and circuit redundancy, *IEEE Trans. on Semiconductor Manufacturing*, 3 No. 3 (1990) 116–127.

[4.5] C.H. Stapper, Yield statistics for large-area IC's, *Proc. ISSCC*, Feb. (1986) 168–169.

[4.6] S.D. Senturia, Microsensors vs. IC's: A study in contrasts, *IEEE Circuits and Devices*, 6 No. 11 (1990) 20–27.

[4.7] R.W. Bogue, Silicon sensors in Western Europe, *Proc. Sensors VI*, IOP Publishing 1993, pp. 119–127.

[4.8] J.B. Sampsell, The digital micromirror device and its application to projection displays, *Proc. Transducers 93*, Yokohama, Japan, 7–10 June, 1993, pp. 24–27.

[4.9] P. Schubert and G.W. Neudeck, Confined lateral selective epitaxial growth, *IEEE Electron Device Letters*, 11 No. 5 (1990) 181–183.

[4.10] M. Bartek, P. Gennissen, E.J. Blaauw, P.M. Sarro, P.J. French and R.F. Wolffenbuttel, A novel silicon colour sensor using SEG, *Proc. Transducers 93*, Yokohama, Japan, 7–10 June, 1993, pp. 144–147.

[4.11] J.J. Pak, A.E. Kabir, G.W. Neudeck, J.H. Logsdon, D.R. deRoo and S.E. Staller, A micromachining technique to a thin silicon membrane using merged ELO of silicon and SiO_2 for an etch stop, *Proc. Transducers 91*, San Francisco, CA, USA, 24–28 June, 1993, pp. 1028–1031.

[4.12] L. Jastrzebski, Comparison of different SOI technologies: Assets and liabilities, *RCA Review*, 44 (1983) 250–269.

5

ON-CHIP FUNCTIONAL INTEGRATION IN MICROMECHANICAL SENSORS

5.1 Introduction

Despite a very long history of impressive IC technology developments, only about 22% of all currently manufactured pressure sensors, and 2% of all acceleration sensors, have integrated passive or active electronics. Most of the pressure and acceleration sensors (the largest sensor market applications) still primarily utilize only the excellent mechanical properties of silicon. The only electronic components in these devices are sensing elements, such as ion-implanted strain-sensitive resistors.

The main reason behind limited on-sensor-chip integration is the highly diversified and limited size of the sensor market. Sensor technology developments were often undertaken by small companies that did not have semiconductor manufacturing technology. The very few sensor gurus who developed successful commercial sensors thus lacked access to state-of-the-art IC technology. IC technologists involved in sensor technology, on the other hand, lacked sensor know-how, delivering highly integrated devices that were not delivering expected performance. Larger semiconductor companies got involved in the silicon pressure sensor technology development early, but later abandoned it as the market development significantly lagged behind the mainstream IC industry (e.g. Fairchild, National Semiconductor, Burr-Brown). Others restricted their efforts to market niches supporting internal higher value added products (e.g. Philips and Texas Instruments).

Only recently have sensors with on-chip electronics started their journey to the fast growing market. The automotive market with its large volumes became the main technology driver, demanding IC pricing for sensors and justifying — with volume orders — the advanced sensor technology developments. The first on-chip integrated pressure and acceleration sensors addressing the automotive market have reached the production phase.

This chapter overviews the current status of the on-chip integrated sensor market, focusing on pressure and acceleration sensors, as these benefit from the excellent mechanical properties of silicon. It references selected aspects of the transistor and monolithic IC market evolution, contrasting them with the integrated sensors market development characteristics. It presents arguments both in favor and against the on-sensor-chip electronics integration. It finally presents two case studies from NovaSensor (a 1985 Silicon Valley silicon micromachining start-up company) development, in which hybrid designs were selected over the single chip solution.

5.2 Why silicon for micromechanics?

Silicon is widely used in manufacturing of electronic integrated circuits where only its electrical properties are used. Thus on the surface, the most attractive aspect of the silicon micromachining technology should be the potential for the on-sensor-chip electronics integration. This potential has not yet been greatly utilized so far, as discussed in the next section. The largest silicon micromachining applications to date, pressure and acceleration sensors, have been enabled primarily by two factors:

- Excellent mechanical performance of silicon enabling it effectively to replace a majority of all other sensing technologies

- The existing infrastructure of the mainstream IC industry, enabling development of products offering an unmatched price-to-performance ratio

While electronics integration has a very high visibility, the on-chip integration of mechanics has a much higher potential impact on volume applications. Integration of the electronics in sensor applications replaces the two batch-manufactured IC chips, sensor and signal processing ICs, with a single batch-manufactured chip. Integration of mechanics, however, replaces a batch-manufactured sensor IC chip and mechanical components manufactured in a *one-at-a-time* mode with a single batch-manufactured chip. The impact of such a dramatic manufacturing technology change on the device cost and volume manufacturing capabilities is quite dramatic. It is far greater than the replacement of the two IC chips with one offered by the electronics-only integration.

Micromechanics may be treated as another dimension of the on-sensor-chip integration: mechanical as opposed to electronic . Similarly, two levels of mechanics-on-a-chip integration are possible:

- Passive elements (e.g. membranes or fuel injection nozzles)
- Active, or microactuator, elements (e.g. pressure valves or motors)

The feasibility of mechanics on chip integration results from the two factors that enabled development of the silicon sensor market: excellent silicon mechanical performance and the IC industry infrastructure. MEMS is also facing a lack of support from the IC industry in selected areas. These factors are briefly characterized below.

5.2.1 Mechanical performance of silicon

Silicon and its derivatives (silicon oxide, nitrite, etc.) are some of the best electrically characterized materials in the world. Their mechanical performances, however, are not well known in the IC industry, and many parameters have yet to be characterized as a function of the process used.

Based on known characterization of silicon, and based on the indirect measurements resulting from the device characterization,

Table 5.1 Comparison of selected properties of steel and silicon

Parameter	Steel	Silicon	Units
Yield strength	4.2 max	7.0	10^{10} dyne cm^{-2}
Young's modulus	2.1	1.9	10^{12} dyne cm^{-2}
Density	7.9	2.3	g cm^{-3}
Thermal conductivity	0.97	1.57	W cm^{-1} $°C^{-1}$

silicon can be classified as the best material for mechanical sensors. Major advantages include high strength, lack of hysteresis and high sensitivity to stress.

(a) Mechanical strength

Silicon strength is comparable to steel, but at a lower density (yielding better dynamic devices depending on the weight to strength ratio) and better thermal conductivity (Table 5.1).

(b) Lack of mechanical and thermal hysteresis, and long-term drift

For all mechanical sensors the measure of excellence and performance limit is defined by the achievable mechanical and thermal hysteresis, and long-term stability. These parameters define the change of the sensor output after half and full mechanical (e.g. pressure or acceleration) and temperature cycles, respectively. Conventional transducer technologies involve many classified 'tricks' as the properties affecting hysteresis are not easily understood and controllable. For example, one of the high-accuracy transducer companies buys tonnes of custom steel, machines sample sensors, and tests long-term stability and hysteresis. If the material is bad, the company orders another lot of steel. All this happens years before using the material on the production floor.

Silicon delivers good performance on each low-cost wafer, thanks to its extremely pure, defect-free crystalline structure.

(c) High sensitivity to stress

The piezoresistive effect in silicon has a stress sensitivity two orders of magnitude larger than that of metal strain gauge, which enables fabrication of the high output devices with simple electronics.

(d) Batch manufactureability

The capability for manufacturing completed mechanical structures simultaneously on multiple wafers, each carrying multiple devices, forms the revolutionary aspect of the silicon micromachining technology: batch manufactureability. None of the other sensor technologies has such an advantage that delivers not only a very low unit cost, but also a very high volume manufacturing capacity.

5.2.2 IC industry infrastructure

Besides the excellent mechanical performance, silicon brings significant support from the established mainstream electronic industry, specifically:

- Access to ultra-pure materials
- Access to advanced semiconductor processes
- Availability of the high volume packaging technologies
- Developed high volume batch manufacturing processes for silicon wafers
- Access to high volume manufacturing equipment
- An available base of educated silicon processing technologists

All these benefits can be 'borrowed' for sensors from the semiconductor industry. The fantastic leverage that the silicon sensor technology enjoys becomes clear when one compares the total R&D funding in sensor and IC technologies. With the exception of solid-state sensors, the total world sensor R&D budget for hundreds of technologies is below one billion US dollars per year. A single integrated circuit technology spends billions of dollars yearly on R&D in materials, processes, manufacturing equipment and training. Silicon-based sensors and microstructures benefit from these vast resources.

Using materials, processes and equipment developed for the mainstream electronic industry significantly brings down the development time and cost for mechanical devices. Silicon micromachining companies are built on the existing high-level infrastructure of the IC industry.

5.2.3 Lack of support from the IC Industry

While mainstream electronics brings substantial advantages to the sensor industry, several aspects of sensor-related technology are unique to the sensor industry and create a substantial technological barrier to entry. These barriers slowed the development of silicon micromachining markets. The required micromachining technologies (in a debugged manufacturing format) include the following:

- Deep silicon etch
- Dimensional control of silicon structures
- Ppm level stability of silicon resistors
- Wafer lamination
- High-volume, low-cost pressure/acceleration/temperature testing
- Chip package stress isolation
- Packaging enabling mechanical interfacing with electromechanical chips

Progress in these fields is developed primarily by the silicon micromachining companies with limited R&D budgets. It is thus not surprising that the introduction of developments is this field is slower when compared to developments in the IC industry.

5.3 Integrated sensor market overview

Pressure and acceleration sensors represent the only visible market for on-chip integrated mechanical sensing devices. This market include two generic levels of electronics integration on sensor chips:

- Passive elements (mainly compensation resistors)
- Active semiconductor elements

Because of limited high-volume sensor applications, the on-chip integration has not produced, so far, the expected return-on-investment (ROI) to justify a faster industrial implementation

Table 5.2 Snapshot of the 1994 sensor market

Item	Sensor Market (million units)			
	Pressure		Acceleration	
Total market	50	100%	6.0	100%
Total silicon sensor market	45	90%	5.8	97%
Silicon sensors with passive integration	7	14%	0.0	0%
Silicon sensors with active integration	4	8%	0.2	3%

of single-chip sensors. The status of the 1994 sensor market is characterized in Table 5.2.

As shown, passive on-sensor chip integration represents only about 14% of the pressure sensor market. Active integration represents about 8% of the pressure sensor market and about 3% of the acceleration sensor market. A limited market share of integrated sensors will likely change in the future, as several large-volume applications are visibly entering the automotive market.

Actual implementations can be grouped into several functional categories. These categories can be summarized as follows:

- Passive temperature compensation and calibration
- Analog electronics for amplification and analog (e.g. laser trimmed resistors) calibration and compensation
- Analog electronics with digital trimming (e.g. with EEPROM)
- Digital electronics with digital trimming
- Advanced electronics, e.g. bus output and/or control algorithm processor (e.g. PID or fuzzy logic) and/or actuator (e.g. power controller)

5.3.1 Passive temperature compensation and calibration

Passive compensation and calibration utilizes primarily on-chip laser trimmed thin-film resistors, fuses and anti-fuses to eliminate sensor errors. The first production of passive temperature compensation

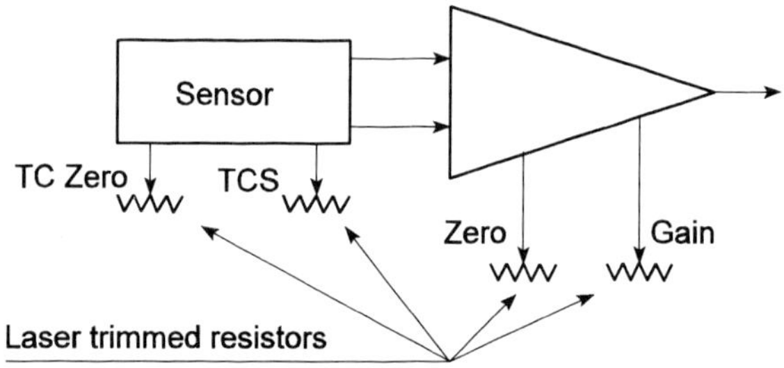

Figure 5.1. *Principle of analog amplification with analog trimming. Typically, temperature errors are compensated at the sensor level, and calibration performed in the amplifier, all with laser-trimmed resistors.*

was implemented by Kulite Semiconductor, USA, in 1979 [5.1]. The first passive temperature compensation *and* calibration was introduced into production by Motorola, USA, in 1986, for disposable blood pressure sensors [5.2]. NovaSensor, USA, introduced a smaller version of such a sensor in 1988 [5.3], but later decided to implement a hybrid version of a transducer, which provided lower cost and higher yields [5.4]. This design is characterized in the case study in section 5.8.

5.3.2 Amplification and analog calibration and compensation

Basic signal conditioning for transducers includes amplification, calibration and temperature compensation. With an analog approach, all these functions are accomplished with analog circuitry: an analog amplifier and laser-trimmed resistors for calibration and compensation (Fig. 5.1).

The first silicon pressure sensors with on-chip amplification were developed in the late 1970s by IC Transducers (first sensor company in Silicon Valley, formed as a spin-off from Fairchild Semiconductor, USA). These integrated sensors delivered such a low yield at that time, that they were not practical for commercial applications.

The second major effort was undertaken by Honeywell, USA. This effort delivered an IPT sensor in 1981, which integrated 3 bipolar operational amplifiers and 8 thin-film laser-trimmable resistors on a single sensor chip. This product delivered excellent performance, meeting requirements of the aerospace applications, but at a cost significantly higher than the alternative implementations. The production program was suspended.

The next major effort to introduce a fully integrated high-level output transducer was undertaken by Motorola, USA. It resulted in the 1985 introduction of the integrated sensor with a 3 V output FSO [5.5]. The sensor was based on a bipolar IC technology, SiCr thin resistors and shear gauge piezoresistive sensing elements. Technology problems delayed manufacturing of this part until 1993, when Motorola started production sampling.

In the meantime Nippon Denso, Japan, made the decision to introduce a monolithic version of the manifold absolute pressure sensor (MAP) for automotive applications. The product was introduced into production in 1989, and replaced a hybrid version of the MAP manufactured previously. As in the other approaches, the sensor design was based on a bipolar process amplifiers and thin-film laser-trimmed resistors, processing a signal from the piezoresistive bridge. Nippon Denso expected that cost of the fully integrated transducer would be higher than the hybrid for the first several years of production, running at about 4 million units per year.

The most recent introduction of integrated pressure sensor came in 1993 from Honeywell to selected larger-volume customers. This chip was an enhanced descendant of the IPT transducer, with improved manufactureability.

In the acceleration field, Nippon Denso developed the first on-chip integrated sensor in 1989. After a several years struggling with technical problems, the sensor moved into 1994 model year production *without* the on-chip electronics. Analog Devices introduced a single chip design for crash airbags in 1991. This design was based on a capacitively polysilicon sensor. On-chip electronics included a servo loop maintaining a constant position of the laterally moving polysilicon fingers [5.6]. The sensor was implemented in a surface micromachined technology utilizing 6 masks and the electronics was designed in a BiCMOS technology that required 22 masks, making a total of 26 masks (two masking steps were common). Limited production shipments started in 1993 after an extended debugging cycle.

Table 5.3 Sensor developments with integrated analog electronics

Company	Type	Trim	R&D	Sales
ICT, USA	P*	No	1977	Never
Honeywell, USA	P	A†	1981	Never
TDI, USA	P	A	1985	Never
Motorola	P	A	1985	1993
Nippon Denso, Japan	P	A	1988	1990
Honeywell, USA	P	A	1992	1993
Bosch, Germany	P	D‡	1993	1994
Nippon Denso, Japan	A§	A	1989	Never
Analog Devices, USA	A	A	1991	1994

* Pressure
† Analog
‡ Digital
§ Acceleration

Chronological industrial efforts in this field are summarized in Table 5.3. As it can be seen, production debugging of the integrated sensor and electronics was a major development theme. It eliminated or significantly delayed production implementation for most of the single chip integrated sensors. All the developments so far have used analog electronics, and only one device used digital trimming technology, with others using analog trimming.

5.3.3 Analog electronics with digital trimming

The major disadvantages of analog trimming are either a low productivity, if potentiometers are used, or substantial investment in process and automated laser trimmers, if a sophisticated thin-film technology is used. Analog electronics with digital trimming combines the best of analog signal processing and its lack of signal quantization, with ease of use and low-cost of digital calibration. It is often called Analog Sensor Signal Processor (ASSP).

Advanced analog signal processing has already entered the transducer field [5.7]. In this approach, an amplifier with digitally programmable gain and offset is used to calibrate and compensate

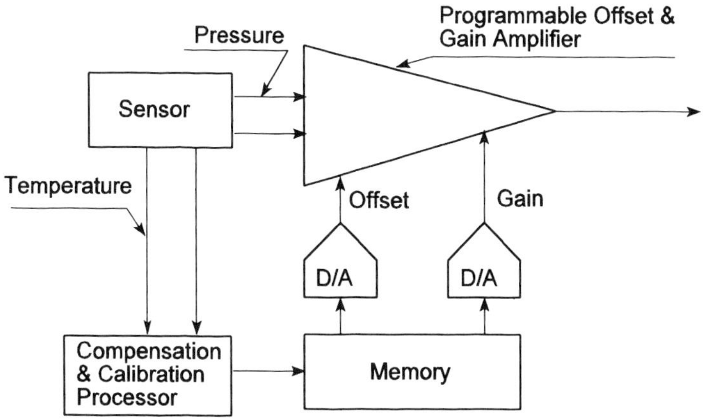

Figure 5.2. *Principle of ASSP. The compensation and calibration processor sets offset and gain of the instrumentation amplifier to compensate sensor errors. Digital memory chip replaces analog potentiometers or resistors. D/A converters are used to convert digital world to analog domain.*

the sensor. Individual compensation and calibration constants are stored in a non-volatile memory. Any type of digital memory can be implemented, and actual smart transducer implementations have used battery-backed RAM, EPROM, EEPROM and several modifications of PROMs. D/A converters are usually used to convert a digital code into an analog control signal for either offset or gain adjustment (Fig. 5.2). In this approach the main sensor signal (e.g. pressure) is in an analog domain all the time, without quantization errors. The temperature compensation circuit changes offset and gain in response to temperature, creating an analog correction signal. This correction signal most often is quantified, creating temperature error-correction quantization error. The magnitude of this error is a function of the internal architecture of the compensation and calibration processor.

There have been several developments from leading universities in this field. The most interesting design came from Berlin University [5.8] and Fraunhofer Institute of Technology [5.9]. The circuit was implemented in CMOS technology, and both capacitive and

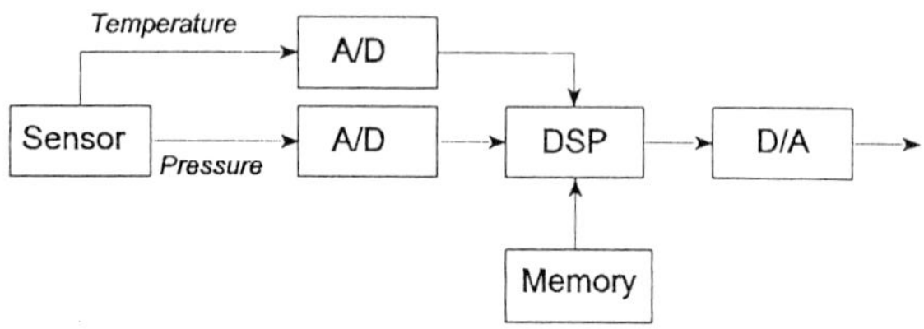

Figure 5.3. *Principle of DSSP. Digitized sensor outputs are processed either by a dedicated Digital Signal Processor (DSP) or microcontroller to provide sensor compensation and calibration. The corrected signal is converted to analog domain, if needed, by the D/A converter.*

piezoresistive sensor approaches were tested. The first commercial implementation of simplified analog electronics with digital trimming was introduced in 1994 by Bosch for the automotive MAP transducer.

5.3.4 Digital electronics with digital trimming

This type of circuitry is often called a DSSP (Digital Sensor Signal Processor) (Fig. 5.3). In this configuration, an analog sensor signal is converted to the digital domain with an A/D converter, compensation and calibration is performed in a digital domain, and the compensated and calibrated signal is converted back to an analog domain.

This approach was used in the first implementations of smart sensors [5.10]. Currently there are several commercial smart sensor designs using this approach; however, none were converted to a single chip solution. The most accurate implementations use 24-bit input A/D converters. This configuration gives the best system flexibility, as it is almost entirely based on the implemented software.

5.3.5 More advanced electronics

There are several advanced system-on-sensor-chip developments at the academic level. Some already have yielded interesting results. These programs include bus output, control processors and actuators. Some of the leading centers in this field are Delft University [5.11], Berlin University [5.8], Fraunhofer Institute of

Technology [5.9], UC California at Berkeley [5.12] and University of Michigan [5.13]. This emerging, high-level integration electronics does not yet have the high-volume market demand to accelerate production implementations.

5.4 Integrated circuit market analogy

The history of semiconductor technology only dates back to the 1950s. Since its invention, transistor technology has evolved from discrete components to integrated circuits with millions of transistors on a single silicon chip. The evolution of the market for some functional modules, such as operational amplifiers, shows some parallelism to the silicon sensor market development.

Initially, operational amplifiers were built using discrete transistors. Eventually, the ceramic-based thick-film hybrid technology was used as a packaging foundation. Thick film resistors allowed efficient laser trimming of functional parameters. Many families of hybrid ICs were built providing performance and reliability, but not necessarily offering the lowest cost for the highest volume. The continuous evolution of the analog IC technology gradually allowed it to enter the applications dominated by the analog hybrids, replacing a majority of them. As soon as more advanced performance becomes available in a monolithic form, equivalent hybrid circuits begin losing market share.

Cost, performance and reliability are the strongest drivers in much of modern commercial electronics. In some areas, particularly aerospace, weight and size are also import drivers for the application development. The same is true for silicon sensors, as many of the techniques associated with silicon micromachining are common with, or adapted from, the IC industry. This includes photolithography, etching, growth of new layers, diffusion and ion implantation.

Since the cost is most often the overriding concern for new applications, enabling or disabling them, it is quite interesting to evaluate the initial cost curves for both transistors and ICs [5.14–5.15] as related technology. Cost learning curves were developed as a function of the cumulative industry volume. For transistors (Fig. 5.4), the cost learning curve exhibits two distinctively different phases. Until the cumulative volume reached 10 million units (1954–

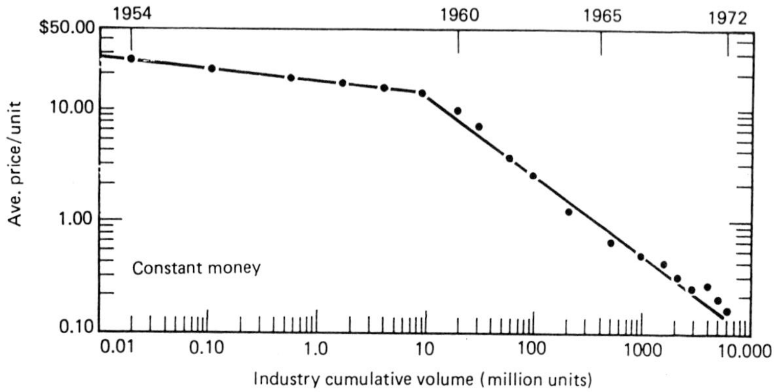

Figure 5.4. *Price experience curve for silicon transistors.*

1960), yearly price reduction was relatively slow. This period of time can be characterized as the basic technology and market development. The rapid explosion of the market within the next 12 years increased the cumulative volume about 600 times to 6 billion units, for a compounded growth rate of 70%/year. Simultaneously, the price was reduced 200 times.

Integrated circuits (Fig. 5.5) exhibited no early price stability. As the basic market was already initiated by the introduction of transistors, integrated circuits were replacing the functional transistor-based assemblies, creating a continuous transition of the market. Cumulative volume increased two times faster than for transistors. The price experience slope was at 73%, indicating a reduction of price by 27% for each doubling of the shipped volume.

The situation for IC sensors is different. While the first silicon-based pressure sensors were developed in 1960, the first mechanically monolithic sensors (a batch-fabricated diaphragm integrated with constraint and strain gages, but without active electronics) were offered in 1967. This sensor level could be compared to a single transistor. Its introduction was about 14 years after the introduction of transistors, and 3 years after the introduction of integrated

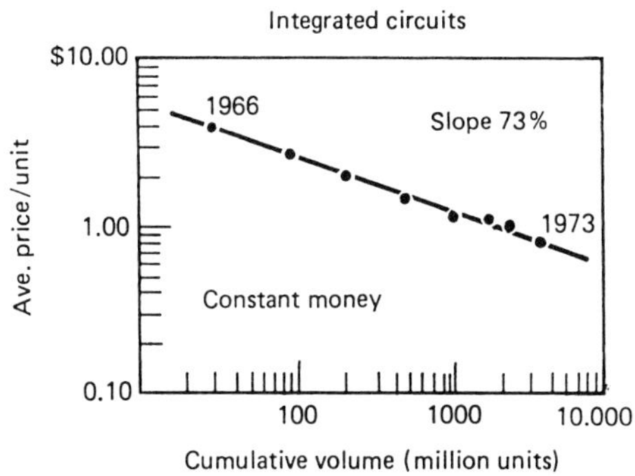

Figure 5.5. *Price experience curve for integrated circuits.*

circuits. The cumulative volume of silicon pressure sensors just recently (in 1993) exceeded the 100 million units. This matched the 1963 status of the transistor industry, or the 1968 status of the IC industry. Transistors needed 10 years to reach 100 million units cumulative volume, integrated circuits only 3 years, but pressure sensors required a full 26 years.

A slower market development created a major economic barrier for the implementation of single-chip integrated sensors. Another barrier was created by the evolution of the applications themselves. Out of the annual volume of 45 million silicon pressure sensors manufactured in 1994, only about 25 million require signal conditioning (the rest interface with the instrumentation designed for low-level output sensors, such as hospital monitors).

A large fraction of the signal conditioning sensor market, about 22 million units, represents MAP transducers manufactured by about ten companies. The largest one is GM Delco, shipping over 10 million MAP transducers per year. Even at this volume it was not cost efficient for Delco to develop a single-chip sensor. Development of the silicon pressure sensor market is illustrated in Fig. 5.6. The price experience curve does not exhibit the initial price-holding period, visible for transistors. It is more similar to the price experience curve for integrated circuits; however, it started from a significantly higher price, and it required three decades to reach the 100 million units which IC s reached in just three years. The average

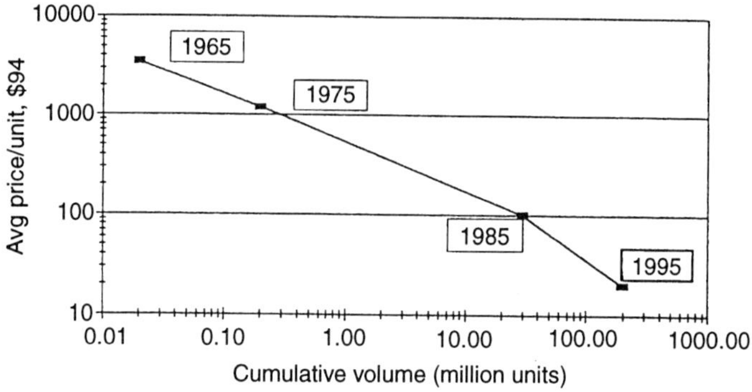

Figure 5.6. *Price experience curve for silicon pressure sensors based on the author's market research.*

price experience slope through 1985 was about 70%, indicating a 30% price drop for each doubling of the cumulative volume. This slope was close to those of transistors and ICs. Since 1985 the price experience curve dropped to 20% (80% price drop for each doubling of the volume), reflecting the rapidly growing low-cost medical and automotive applications. This slope was significantly steeper than those for transistors and integrated circuits; however, the pressure sensor market growth was slower.

Development of the acceleration market was even slower than the pressure sensor market. The first sensor was demonstrated in 1967. The cumulative volume of silicon acceleration sensors in 1993 was below 250,000 units. In 1994 the market jumped by about 5 million units. It is forecast that a cumulative volume of 100 million units will be reached by the year 2000.

It is clear that while the roots of the silicon sensor market are the same as for other semiconductor devices, the dynamics of this market are quite different. These dynamics are closer to the development of markets in which sensors, currently the only large-volume micromachined product, are used. These markets include the following segments: process control, industrial, military, aerospace, commercial aviation, medical, automotive and consumer. Within

Table 5.4 Market development for transistors, integrated circuits and silicon sensors

Item	Transistors	ICs	Pressure sensors	Acceleration sensors
First products	1953	1965	1960	1967
Cum 100M units	1963	1968	1993	2000?
Years to reach 100M	10	3	33	33?
1994 volume, units	Huge	Huge	45M	6M

these segments, only the automotive and consumer markets, currently estimated respectively at 300M and 50M, have the potential to create growth rates somewhat similar to the IC market growth.

5.5 Monolithic or hybrid

Hybrid technology is typically based on the ceramic substrate, which is used as a packaging base for monolithic IC chips and other components. The same ceramic substrate is also used as a base for the fabrication of the interconnecting metallization and thick-film resistors, which are typically screen printed in an array form at room temperature, and fired at 850°C. Ceramics have good mechanical properties and offer easy customization at low non-recurring engineering cost (in comparison with custom monolithic ICs). Development cycle times are short, and volume production capabilities are high.

Hybrid integrated circuits have been in use since the 1950s, and are still widely used. This technology offers several advantages for many sensors:

- Simplified pressure connectivity, as many pressure port configurations can be directly integrated into the package

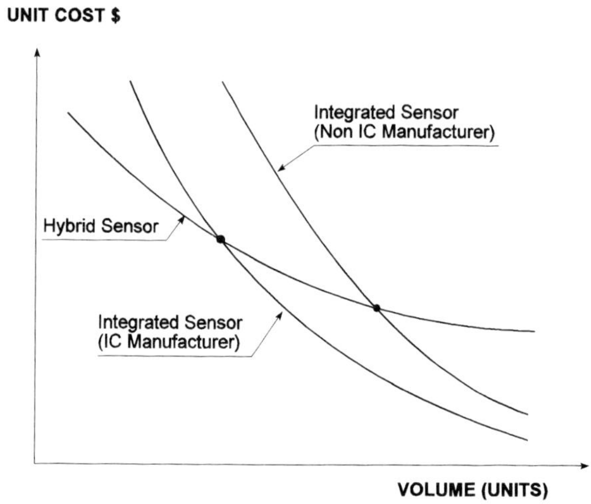

Figure 5.7. *Hybrid sensors are less expensive in smaller volume. The IC manufacturer will require a smaller volume to match the cost of hybrids with a monolithic solution, as compared to a non-IC manufacturer.*

- Capability for a high-density packaging of multiple IC chips on a single substrate
- Access to low-cost laser-trimmable resistors
- Low-cost hermetic feedthroughs from the pressure cavity (vias)
- Broad equipment and process base developed by the electronics industry
- Capability for handling multiple sensors on a single ceramic plate (snapstrate) through the entire production process, enabling a batch sensor assembly process, equivalent to batch wafer processing
- High reliability proven in both automotive and aerospace applications

Similarly to the hybrid analog modules, as long as the production volume is below a critical size, the hybrid design is less expensive (Fig. 5.7). This critical volume varies significantly as a function of the existing company infrastructure. For semiconductor companies, the monolithic design delivers a lower manufacturing cost at lower volume, typically around 1–3 million units/year. For more transducer and system companies, the critical volume is typically between 5–20 million units/year.

As mentioned in chapter 4, the wafer processing facility has a high fixed cost. The small volume processing cost (e.g. one wafer/month) is practically the same as the processing cost of several hundred wafers. More importantly, the wafer fabrication yield increases significantly with increased production volume. Each wafer typically carries between 100 and 10,000 sensors. A high production volume for fully integrated sensors is thus a mandatory requirement to make them cost competitive with the hybrid implementations. Due to a high volume of non-micromachined wafers processed in the IC-manufacturer facility, cost allocation for micromachined wafers is lower, and their yields higher, as compared to the non-IC-manufacturer.

Interestingly, when the IC technology brings new generations of products, hybrid technology can assembly them on a single substrate for an even higher level of integration. The newest trend, Multi-Chip-Modules, is one of the hottest new hybrid technologies. It is thus expected that a similar evolution will be visible in the sensor industry, and hybrid sensors will co-exist with monolithic designs for a long time.

5.6 Integrate or not to integrate circuits with micromechanical sensors?

There are many compelling reasons behind the drive towards single chip signal conditioned sensors. Probably the most attractive one is the past track record of the IC industry. Single chip circuit integration delivered an unprecedented price-to-performance ratio in almost every application, having no competition from any other technology. The sensor industry expects the same advantages from the single chip sensors.

The silicon sensor industry foundation is created by the adopted IC industry technologies. There are many aspects of sensors, however, that must be developed by the sensor companies on their own. These sensor-specific developments are related to such issues as:

- Deep wafer etching
- Compatibility between micromachining and IC processes
- Package stress isolation
- Mechanical interfacing with a real world (e.g. providing media compatibility, threaded pressure ports, static pressure package withstand)
- Parametric testing, e.g. in response to pressure and temperature
- Requirement to deliver specific performance, such as pressure nonlinearity and many others.

The sensor industry often demands from silicon designs performance not needed nor understood in the IC world, such as ppm level silicon resistor stability, low mechanical and thermal hysteresis, stress free processes, etc. On-chip electronics integration thus requires a significant investment by the sensor companies.

Once a decision to integrate the on-sensor-chip electronics has been made, two major issues have to be resolved:

- *Logistics of pressure or acceleration testing and trimming*
  In discrete sensor applications, performance of the sensor can be modelled without the influence of the signal conditioning circuitry. Once the electronics is permanently connected, it is more difficult to perform temperature compensation. Thus the logistics of testing and trimming could be significantly more complex, as compared to discrete sensors.
- *Possible degradation of sensor performance*
  Most pressure and acceleration sensors use piezoresistive sensing elements. These elements are sensitive to all parasitic stress. This stress can be (and will be) generated by the additional materials and structures integrated on the chip, each having a different coefficient of temperature expansion. Moreover, the additional IC processes may alter the basic properties of the sensing elements, making compensation more difficult. Taking this into account, the design and debugging process for fully integrated sensors is significantly longer than for discrete devices.

The silicon sensor industry is still small. Sensor-related know-how resides within a limited number of sensor gurus, most outside of companies that can afford an advanced on-chip integration. It is thus not surprising that the learning curves for single chip sensors were (and are) painfully slow.

There are many pro-integration arguments, as there are many against. These have been discussed in general in the previous chapters and are discussed in more detail for the micromechanical sensor.

5.6.1 Pro-integration arguments

(a) Size and weight

Many of the new applications require that a sensor be comparable in size and weight to other integrated components. The smaller the sensor, the more functional the final product can be made (e.g. a barometer in the wrist watch). Some other applications can even offer cost savings when the weight is reduced. For example, in modern aerospace systems, the cost of adding weight to an aircraft can be quantified and is not insignificant even for items such as sensors. Increasingly, the same is true for automotive and industrial sensors. External dimensions are also becoming critical. Even using small outline (SO) packages, the effect of replacing two individually packaged components with a single integrated component reduces the system weight and space by a factor of at least 2 or 3.

(b) Reliability

Integration has a major impact on reliability simply because of the reduction in the number of components. This includes the active components and also wire links, circuit substrates, and the numerous electrical joints otherwise to be found between the sensor and its signal processing. Other changes that may have a noticeable effect are the reduction in electrical power consumption which typically accompanies the integration of multiple components into a single part.

(c) Cost in volume

Once the initial development and engineering cost of an integrated sensor has been met, the manufacturing cost can be expected to be significantly lower than for a multi-component sensor. The most obvious reason is that not only has a number of parts and interconnections been removed from the sensor, but much of the assembly cost of those parts has disappeared with them. A less obvious benefit is that much of the testing can be carried out while the sensors are still in wafer form. This creates a major challenge for sensors (e.g. pressure or acceleration sensitivity testing), but once solved, facilitates rapid automatic testing, borrowing techniques from the IC industry. The ability to weed out the majority of failures at the wafer stage before adding the substantial cost of sensor packaging is a substantial manufacturing cost saver.

(d) Temperature variation

Few sensors are free of the burden of cross-sensitivity to temperature and so hard is the burden to avoid, that in most cases it is accepted and correction is attempted in the signal processing. In many cases, however, it is inconvenient to attach a temperature sensor to the basic sensor, and a temperature sensor is often included in the signal conditioning or attached to a part of the package. The question that immediately arises is: is the basic sensor at the same temperature as the temperature sensor? In an integrated sensor, this must be true within close limits as both are part of the same piece of silicon. A further advantage accrues to the integrated sensor if the signal conditioning circuit is also temperature sensitive. Since the circuit and the sensor are inherently inseparable and at the same temperature, circuit temperature compensation can easily be achieved in the same act as sensor compensation and therefore for zero cost.

(e) Feasibility of arrays

The integration of several sensors onto a single piece of silicon is not much more of a challenge than the integration of one. Packaging costs are not likely to be significantly more so the main additional cost is only that of silicon area. With a multiplicity of sensors, statistical comparison of several sensors of the same type becomes possible for diagnostics and calibration monitoring. Alternatively, arrays of sensors with multiple cross-sensitivity can be used to calculate outputs free of cross-sensitivity (e.g. electronic noses).

(f) Reduced parasitics

Because of the small dimensions of an integrated sensor, the parasitic impedance of associated interconnects is dramatically reduced and also because of the dimensional stability arising out of the method of manufacture. This is of immense benefit in the design of capacitive sensors, where, for example, position is converted to a small capacitance change, often significantly smaller than the interconnection capacitance. The capacitance variation which constitutes the signal must frequently be resolved to a few fempto-farads and any variation due to parasitics which is greater than that invalidates the sensor accuracy.

(g) Increased RFI immunity

The small dimensions and short length of the signal paths carrying unamplified signals have an additional benefit. By the crudest possible analysis, a 1 mm connection even in a most demanding automotive requirement of 200 V m^{-1} field could only acquire 200 mV of RFI (actual values would be very much smaller and would depend on impedances and geometry). Should even these low levels present a problem, then of course it is still a lower-cost option to screen a 5 mm cube than, for example, a 25 mm cube.

5.6.2 Counter integration arguments

There are many difficulties in integrating the sensor and electronics and specific issues which might make integration less desirable for a particular application. These issues are characterized below.

(a) Non-recurring engineering (NRE) costs

The design (as opposed to manufacture) of fully integrated sensors is never likely to be cheap. Analog and digital (mixed mode) simulators will be required for the circuit design as well as tools for modelling physical performance of the new sensor. The prospect of allowing circuitry to encroach on areas actively involved in mechanical aspects of the sensing function raises the prospect of as yet undreamed of simulator functions. Nor is the manufacturing plant required to produce integrated sensors likely to be cheap. It is for this reason that designs currently under discussion or recently introduced are all intended to be produced in near-standard IC fabrication facilities.

(b) Manufacturing volume

The high NRE cost discussed in the previous section means that integrated sensors are only likely to be available where that cost can be supported by very-high-volume applications or where a high cost per sensor can be tolerated over moderate volumes. Outside these scenarios, the only hope for low-volume sensors in the foreseeable future is to share NRE costs with a similar high-volume application.

Many sensor manufacturers may elect to consider using an external wafer foundry service to process the IC wafers, and finish micromachining in-house. This approach creates a problem for smaller volumes (by IC standards), as foundries with advanced and stable processes are hesitant to accept orders for volumes smaller than, for example, 500 wafers/year.

(c) Media protection

An accelerometer can be hermetically encapsulated in the medium of the manufacturer's choice — even a vacuum. Pressure, flow and chemical sensors must be in intimate contact with the sensed medium, and require reliable media protection to avoid mechanical or chemical destruction. These are problems to which the sensor is inevitably subjected, but not the signal processing circuit. Typically the circuit would be kept well away from any media, especially one which might carry abrasive particles or unknown chemicals. In the case of the integrated sensor, either both sensor and circuit must be exposed to the medium, or complex arrangements will be required to isolate one part of the silicon surface (with the electronics) from an adjacent part (with the sensor) exposed to the medium. The former arrangement promises severe technical problems while the latter is likely to be expensive in itself and to require larger silicon chips to accommodate the barrier area.

(d) Combined yield

Both sensor and IC go through an increased number of process steps. This increases somewhat the cost of both sensor and signal conditioning, as the section of the silicon chip with the sensor (usually 6 to 12 masking steps) undergoes the unnecessary IC process (13 to 22 masking steps), and vice versa. Usually only the metallization step may be combined for both processes, and all other steps have to be performed.

The major cost problem is related to the reduced overall yield. If both products yield 80%, the combined yield will be only 64%. Unless both sensor and IC design are high yielding, the economic impact of the on-sensor-chip integration may be unacceptable. It is an inevitable fact that for a given semiconductor process, the yield falls exponentially as die area increases. In the case of the integrated sensor, not only will the chips be larger because they contain the sensor and the electronics, but also the process will be more complex having more steps of a wider variety. The more complex process will lead to an additional reduction in yield.

It is also possible that even for devices which nominally work, performance will be degraded as the sensor will be subject to circuit processes which it would not normally see, and likewise the circuit would be subject to sensor processes.

5.7 Mechanical on-sensor-chip integration

The integration of mechanical features on silicon sensor chips provides a potential for substantial benefits to the users, far outweighing the benefits of electronic integration. The existing sensor-related mechanical designs are manufactured using traditional one-piece-at-a-time technology. The integration of mechanical components on a single chip offers a high-volume low-cost batch mode manufacturing technology, which rewrites the economics of many products.

Several examples of the on-sensor chip mechanical integration (or enhanced micromachining content) include the following:

5.7.1 Acceleration sensors

Two major mechanical functions are currently integrated on the sensor chip: shock protection and self-test. Shock protection allows the survival of the sensor during drops on the floor — generating up to 10,000g shock — for devices with an operating range of only

$2g$. In the classical acceleration sensor designs, this function was accomplished by a screw-driven mechanical stop.

The self-test generates a mechanical deflection of the seismic mass, simulating the effect of acceleration. It enables testing the integrity of the sensor structure in critical applications, such as automotive airbags. Several of the new acceleration sensors have the self-test integrated directly on a chip, while the older generation had been using mechanical hammers to perform the same function.

5.7.2 Pressure sensors

Process control pressure transducers require a single-sided pressure overload survival of up to 300 bar for differential pressure sensors measuring just 0.07 bar full scale. Classical designs use expensive mechanical configurations, mostly based on large metal diaphragms activating the valves which seal off the internal flow channels for the silicone oil surrounding the sensor. This mechanical design protects the sensor, however, at a high cost. The newest silicon sensor designs incorporated the overload protection on the sensor chip, dramatically simplifying the transducer design.

5.7.3 Flow controllers

The current generation of products is based on the discrete pressure sensor and valve. There are industrial developments targeting integration of both these components on a single chip, promising a significant reduction of the size and cost.

5.7.4 Microactuators

The current generation of disk drives uses a servo positioner controling the arm which holds the read/write head. Positioning accuracy is limited to several microns, and response time is limited by the arm's compliance. The integration of the microservo positioner directly at the head location enables an improvement of the positioning accuracy to 0.1μm, with a simultaneous reduction of the response time. These improvements together enable the development of hard drives with 25,000 tpi (track per inch) density, supporting 10 GB in^{-2} read/write densities. Such densities were scheduled 10 years from now based on the hard drive industry projections using the classical technology.

These few examples clearly illustrate the potential impact that mechanical on-chip-integration may have on the industry. Mechanical integration could become the fastest growing segment of silicon micromachining, with a larger application base than the electronics integration.

5.8 Case study: hybrid against monolithic

NovaSensor was founded in 1985 in Silicon Valley, California, with the objective of bringing advanced MEMS devices to production. In 1990, the company was acquired by Lucas and the name was changed to Lucas NovaSensor. Many new MEMS devices were developed within the first years of NovaSensor's growth; however, the major focus was on pressure sensors, as they represented the only developed high-volume market segment at that time.

Two of these developments, where the on-sensor-chip integration had to compete with the hybrid solution, are characterized below.

5.8.1 On-sensor-chip passive electronics integration

One of the largest markets for pressure sensors is the disposable blood pressure sensor market. It currently consumes about 17 million pressure sensors yearly. As hospital monitors maintain the transducer sensitivity standards initiated decades ago, including 1% calibration of the 5 μV V^{-1} $mmHg^{-1}$ pressure sensitivity and an option for the AC excitation, sensors in this application cannot use active circuits. The only possibility for on-chip integration is a passive compensation and calibration circuit.

Only silicon piezoresistive sensors are used in these applications. To meet specifications recommended by the US Association for Advancement in Medical Instrumentation (AAMI [5.16]), a *de-facto* world standard, a passive signal conditioning circuit must perform the following functions:

- Compensate sensor's initial offset (zero) to a maximum of ±25 mmHg

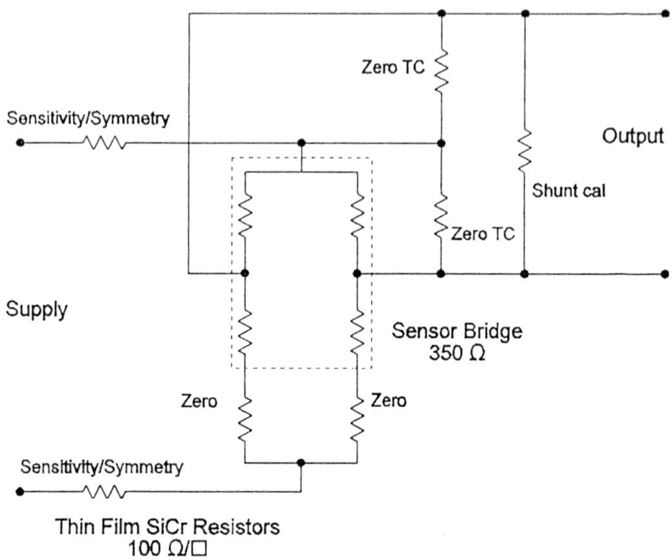

Figure 5.8. *Functional circuit used in passive compensation and calibration of disposable blood pressure sensors.*

- Compensate zero temperature coefficient to a maximum of ±0.3 mmHg $^{\circ}C^{-1}$
- Compensate pressure sensitivity to 5 μV $V^{-1}mmHg^{-1} \pm 1\%$
- Compensate temperature coefficient of pressure sensitivity to below 0.1% $^{\circ}C^{-1}$
- Calibrate the output resistance to either 300 or 350 Ω, to enable the shunt calibration feature built into the monitors
- Provide an output common mode voltage equal to 0.50 ± 0.05 of the supply voltage

The functional circuit delivering all these functions is shown in Fig. 5.8. A low-impedance sensor bridge is used. Several temperature-stable resistors perform compensation and calibration functions, as shown in the drawing. These resistors have to be individually adjusted, or trimmed, for each sensor. The most popular

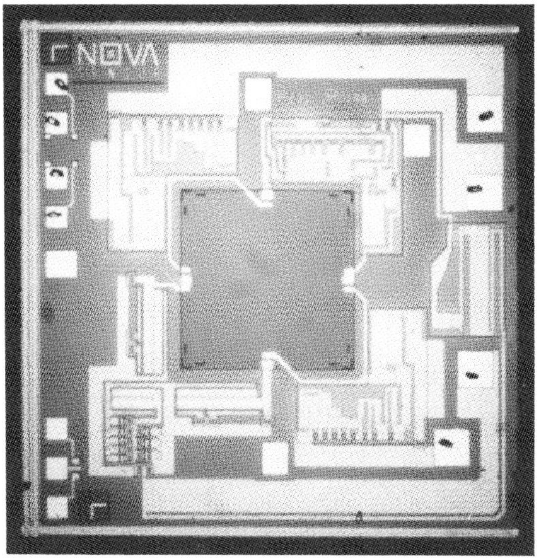

Figure 5.9. *NovaSensor single-chip blood pressure sensor incorporated laser-trimmed SiCr thin-film resistors for compensation and calibration.*

resistor correction technology is laser trimming. Laser trimming can be performed either on thick-film or thin-film resistors.

In 1987 NovaSensor started production of a hybrid disposable blood pressure sensor. A silicon sensor die was mounted on a ceramic substrate with laser-trimmed thick-film resistors. In the same year, the company initiated the development of the next generation of such a sensor based on the on-sensor-chip integrated laser-trimmable SiCr resistor technology. The resulting sensor chip P231 was the smallest on the market at the time, with a size of 2.5 mm^2 (Fig. 5.9). In addition to basic sensing and calibration, the sensor's layout included several test devices, enabling in-process performance monitoring.

Manufacturing started in 1988, and after shipping approximately 100,000 transducers, production was stopped in 1989. The production cost of the on-chip integrated sensor was consistently higher than the cost of the hybrid version of the transducer manufactured in parallel. NovaSensor was a micromachining company, and the cost parity between hybrid and monolithic implementations was at a significantly higher production volume than for an established IC manufacturer, such as Motorola.

The development of the new generation hybrid sensor was initiated. This effort resulted in the development of an advanced manufacturing technology based on a batch processing of snapstrates, ceramic $4.5 \times 4.5''$ (11.4×11.4 cm) plates, each with 120 sensor cells. Each cell had an attached sensor die, a plastic pressure port, and thick-film resistors for calibration and compensation. Throughout the entire manufacturing process snapstrates were handled in cassettes carrying between 12 and 30 plates (depending on sensor design). Most of the assembly equipment (such as die attach, wirebond, etc.) was designed to take the cassette and process all the plates automatically. The line achieved a throughput over 1000 units per hour (uph). Due to a very low production cost and appealing design, production volume increased dramatically, forcing the installation of a second production line.

Fig. 5.10 shows a disposable blood pressure transducer located on the snapstrates. A plastic housing provides a flow-through channel for pressure media (saline solution), as well as a fast flush device enabling removal of air bubbles from the system. A ceramic snapstrate carries 120 sensors. One side of the snapstrate has plastic pressure ports, while the other carries thick-film resistors. Connection between both sides of the ceramics is accomplished with vertical vias.

Snapstrate plates are laser scribed (partially cut). After completing test and assembly operations, ceramic plates are broken, or snapped, along the prescribed lines, yielding compensated and calibrated pressure sensor units.

As a result of batch processing, the standard cost of the sensor is very low, enabling a competitive position on the market against the on-sensor-chip integrated sensors from Motorola.

While batch processing was a new technology for sensors, it was developed quite a long time ago for hybrid integrated circuits. Most of the automated equipment used on Lucas NovaSensor's production line was originally designed for the hybrid ICs.

The advantages of hybrid technology in this application can be summarized as follows:

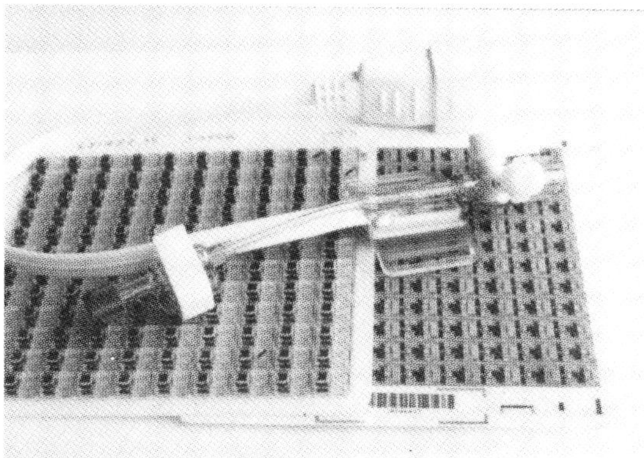

Figure 5.10. *Lucas NovaSensor snapstrate technology mirror batch wafer processing — 120 pressure sensors are located on one ceramic snapstrate, and 12–30 snapstrates are processed in one cassette. One side of the snapstrate includes plastic pressure ports, and the other side laser-trimmable thick-film resistors. Hospital packaging of a hybrid sensor is shown, courtesy of Baxter Healthcare Corporation.*

- Low entry cost
- High volume manufacturing capability resulting from batch manufacturing capability (multiple units on a single plate)
- Fast development cycle
- Mechanically stable packaging base for the sensor die
- Very low cost laser-trimmable thick-film resistors
- Inexpensive technology of hermetic interconnections connections (vias)
- Good available equipment and process base

The problems with single chip sensor technology in this application resulted from the following factors:

- Significant investment, for non-IC manufacturer, in new process development and manufacturing equipment
- Long leadtime for critical equipment delivery (e.g. 9 months for a dedicated automated thin-film laser)

- Long process yield debugging cycle and lower yield as compared to hybrid technology
- Requirements for a micron-level sensor die alignment in the package to enable fast in-package laser trimming, forcing several iterations of the handling equipment design
- Higher effective die cost as compared to a hybrid approach

Once the volume of hybrid sensors approached 10 million units/year, the reintroduction of the single chip integrated blood pressure sensor was considered.

5.8.2 On-sensor-chip active electronics integration

Many applications require signal-conditioned high-level output pressure sensors. In 1988, NovaSensor initiated the development of a single chip pressure sensor to support the microprocessor-based sensor applications. A bipolar IC process and laser-trimmable thin-film SiCr resistors with a bulk micromachined sensor process were selected as a foundation for the sensor.

The functional circuit is shown in Fig. 5.11. A traditional two op-amp instrumentation amplifier configuration was selected for the amplification. A proprietary zero temperature compensation technique was used, and a statistical temperature compensation of the FSO was implemented. Calibration of zero was performed within the sensor bridge, and calibration of FSO was achieved via the amplifier's gain control.

A unique output stage circuit design enabled a rail-to-rail output supporting the automotive transducer requirements, providing either a 0.5 to 4.5 V or 0.25 to 4.75 V output at a single 5 V supply voltage.

Analog laser trimming of thin-film resistors was selected. Special attention has been devoted to the logistics of the test and trim procedures. Discrete sensors allow an easy characterization of their performance. With integrated electronics, the sensor is permanently wired to the electronic circuitry, and many of the standard sensor modelling techniques cannot be used.

The layout of the chip is shown in Fig. 5.12. The chip size was 2.5×2.5 mm^2, designed as a three terminal device. Several

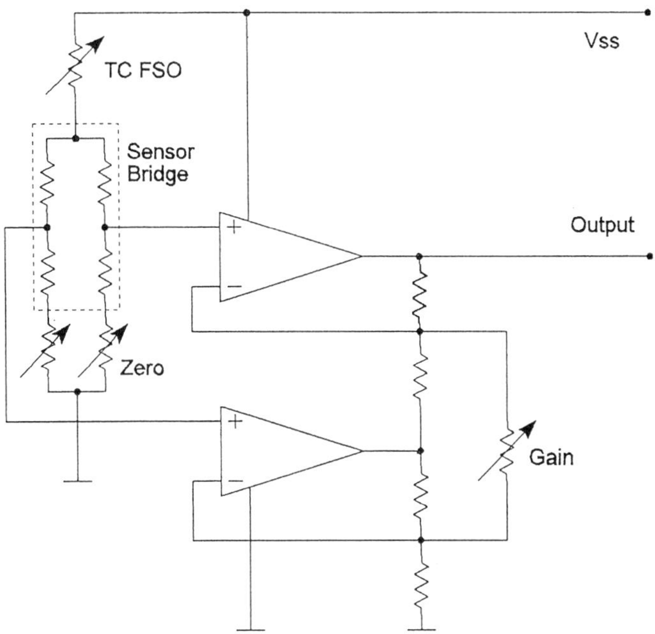

Figure 5.11. *Functional circuit of the single-chip pressure sensor using a bipolar IC process and thin-film SiCr resistors.*

additional pads were included to facilitate access to the internal test points used for performance modelling. The sensor was designed on <100> wafers, and the performance of the bipolar transistors had to be tested and modelled in this orientation. Standard piezoresistor configuration was used.

Facing aggressive market pricing, this sensor was never released to production. Instead, development of an analog hybrid transducer was initiated, followed by the development of the smart hybrid transducer.

To address very low cost high-volume applications, a simple P592 type sensor was selected for the smart transducer. This sensor was fabricated using silicon fusion bonding that yielded the smallest high-volume production pressure sensor chip: only

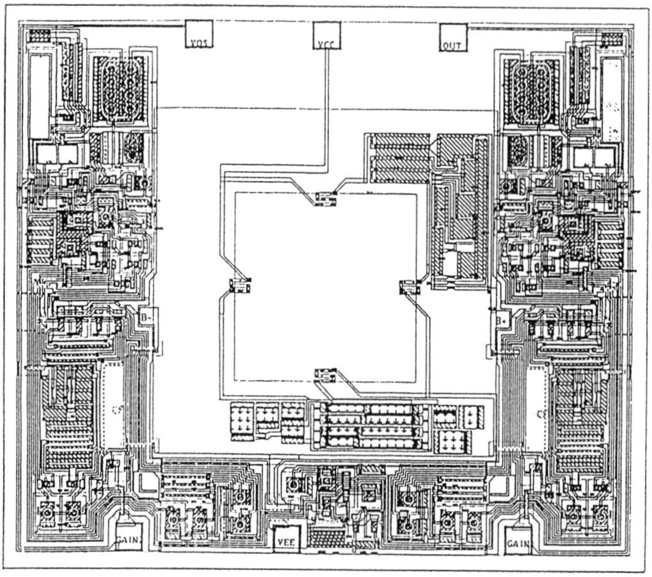

Figure 5.12. *Layout of the single-chip pressure sensor based on a bipolar IC process, SiCr thin-film resistors and bulk piezoresistive micromachining process.*

$1 \times 1 \times 0.6$ mm^3. The chip was processed on 4″ wafers, each carrying 6600 sensors. The chip layout (Fig. 5.13) included only the basic sensor bridge and metallization. Silicon fusion bonding allowed the incorporation of a silicon constraint, which delivered very good rejection of package stress, thus making possible a direct die attach to mechanically unstable materials, such as printed circuit boards. A first generation smart signal conditioning ASIC was developed by a Silicon Valley start-up company, ISS. This ASIC (Fig. 5.14) was developed using a DSSP configuration in a low-cost 3 CMOS technology. It incorporated digital trimming technology. The circuit incorporated a digitally programmable input amplifier, 10-bit A/D converter, dedicated compensation and calibration DSP processor and 9-bit D/A output. Performance of all these functional blocks was programmable from the small 256-bit EEPROM.

The most innovative feature of this ASIC was the implementa-

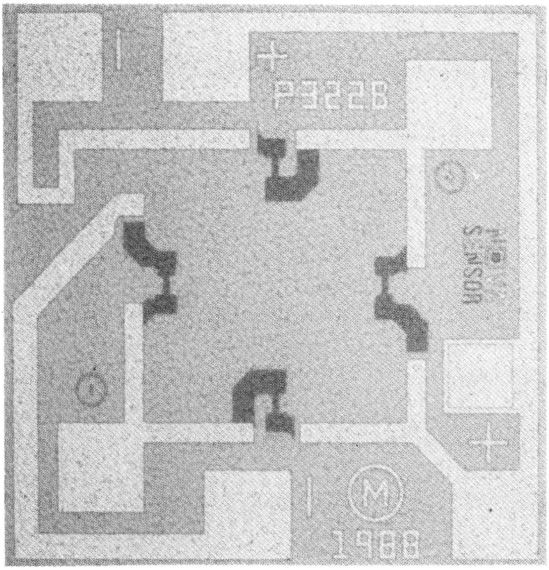

Figure 5.13. *Silicon fusion bonding enabled NovaSensor to develop the world's smallest high-volume production pressure sensor: 1 × 1 mm.*

Figure 5.14. *Lucas NovaSensor's pressure transducer used ASIC with built-in local bus, enabling batch manufacturing of transducers.*

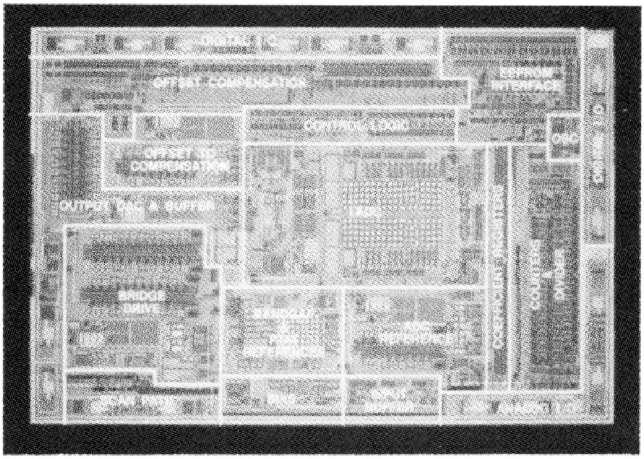

Figure 5.15. *Smart signal conditioning VLSI ASIC implemented in 3 μm CMOS was too complex for on-sensor integration.*

tion of a proprietary local bus. This bus enabled a parallel connection of multiple sensors in the test oven, and the transmission of analog and digital data from each sensor at each test pressure and temperature to the test computer. After calculating the correction coefficients, the test computer was downloading compensation/calibration directly to the transducers' EEPROMs. In a consecutive operation, the final test was performed.

To take advantage of the ASIC's bus capabilities, the transducer subassembly was designed on a large printed circuit board, called a 'snapboard'. Each snapboard carried multiple transducer assemblies, similar to the ceramic snapstrate technology, or the silicon wafers. A four-layer snapboard was terminated with a connector. Several snapboards were inserted into the test oven, enabling very efficient high-volume testing. After completing the final test, the transducers were separated into individual units, and packaged into the plastic housing (Fig. 5.14).

The incorporation of the manufacturing bus enabled the integration of three operations, pre-test, trimming and final test, into one manufacturing operation with three automated steps. It translated into a lower manufacturing cost and higher yield, as only one pressure connection to the transducer had to be made on the production line.

Integration of the smart and advanced features on a sensor chip would be too complex to be debugged at a reasonable cost in a short time frame. The separation of sensing and signal processing between two expert companies, two wafer fabrication facilities and two silicon chips yielded a solution, efficient in cost and development time.

5.8.3 Summary

In both the examples discussed, hybrid technology was found to be more cost advantageous than the equivalent monolithic, on-sensor-chip integrated solutions. In the case of medical disposable sensors this was true even for a very large volume production, in excess of 5 million units/year.

5.9 Conclusions

In the IC industry, a higher level of integration has enabled the reduction of the number of components in existing designs, immediately bringing economic benefits. The size of the market was large enough to justify a quick return on investment for a higher level of integration.

The sensor industry was smaller than the IC industry, and much more fragmented. For years there were only two multimillion unit/year applications: disposable blood pressure and manifold automotive pressure. This application has started to attract single chip sensors; however, their cost was comparable with or higher than that of the hybrid designs.

The sensor market is in the process of very rapid growth. Over the next decade, the pressure sensor market is expected to grow from its current (1994) 50 million units (45 million based on silicon micromachining) to about 150 million. The acceleration sensor market will grow from its current 6 million to an expected 35 million units. These volumes justify the development of fully integrated devices. The largest market will be the automotive segment, clearly driving the progress in sensor technology. The first fully integrated devices have already entered the market place. In automotive volumes, these sensors may be more cost effective than the hybrid technologies. To be competitive, all key automotive market players will have to introduce advanced sensor technology; it is not clear, however, if a monolithic approach will be the winner. Hybrid approaches still have competitive advantages.

As was shown in the case study, hybrid technology has been a very good solution for pressure sensor applications. It may be expected that hybrid designs will be used in many new products. It should be noted that many advantages of hybrid technology result from the special packaging needs of MEMS, which do not exist in electronic IC business.

While several different IC technologies can be used for on-sensor-chip integration, the most attractive one seems to be the CMOS technology used for high-volume digital integrated circuits. If the sensor circuit is developed in this technology [5.17], than it can benefit from the continuous feature size reduction already 'pre-scheduled' for the next decade, thus reducing the cost.

To take full advantage of the integrated sensor die size reduction, new sensor technologies should be explored. These technologies should minimize the number of overall masking steps as well as the resulting chip size. One of the re-emerging options is the utilization of the junction-based sensors. These types of devices [5.18–5.20] may utilize the same transistors that are used for signal processing, potentially eliminating several masking steps unique to piezoresistors. Smaller die sizes will require thinner mechanical structures, which may lead to broader utilization of resonating sensing technology [5.21].

The question 'To Integrate or Not to Integrate Sensor Electronics' seems to be clearly evolving into *'How to Integrate Sensor Electronics for Automotive Applications'*.

5.10 References

[5.1] J. Bryzek and K. Peterson, *Silicon Sensors and Microstructures*, NovaSensor, Silicon Valley, September 1991.

[5.2] R. Frank and W. McCulley, An update on the integration of silicon pressure sensors, *Professional Program Session Record 27*, Wescon/85, San Francisco, 19–22 November, 1985.

[5.3] J. Bryzek, New generation of disposable blood pressure sensors, *Proceedings of Sensors Expo*, Detroit, MI, 1987.

[5.4] J. Bryzek and D. Gee, Low-cost high-volume pressure sensors, *Proceedings of Sensors Expo 94*, Cleveland, OH, 18–20 September, 1994.

[5.5] W. McCulley, C. Darlington and J. Gregg, Fully integrated monolithic pressure sensor with on-chip calibration and compensation, *Proceedings of OEM Design Conference*, Philadelphia, 9–11 September, 1985.

[5.6] F. Goodenough, Airbags boom when IC accelerometer sees 50 g, *Electronic Design*, 8 August, 1991.

[5.7] J. Bryzek, Evolution of smart transducers design, *Proceedings of Sensors Expo West*, San Jose, CA, 2–4 March, 1993.

[5.8] E. Obermeier, V. Schlichting, Smart pressure sensor with on-chip calibration and compensation capability, *Proceedings of Sensors Expo West*, Anaheim, CA, 8–10 February, 1994.

[5.9] U. Schöneberg, F.V. Schnatz, W. Brockherde, P. Kopystynski, T. Mehlhorn, E. Obermeier and H. Benzel, Smart CMOS capacitive pressure transducer with on-chip calibration capability, *Sensors and Actuators*, A34 (1992) 77–83.

[5.10] E. Kompass, Smart transmitter stores calibration data, *Control Engineering*, 10/1983.

[5.11] E. J. Hogenbirk, H.-J. Verhoeven and J. H. Huysing, An integrated smart sensor for flow and temperature with I^2C bus interface, *Proc-ISCAS95*, Seattle, WA, USA April 30–May 3 1995, pp. 2225–2228

[5.12] W. Yun and R. Howe, Σ-Δ modulator interfacing with silicon microsensors, *Proceedings of Sensors Expo West*, San Jose, CA, 2–4 March, 1993

[5.13] K. D. Wise, Integrated microinstrmentation systems: smart peripherals for distributed sensing and control, *Digest IEEE Int. Solid-State circuits Conf.*, San Francisco CA USA, Feb. 1993, pp. 126–127.

[5.14] B . Hedley, A fundamental approach to strategy development, *Long Range Planning*, December 1976, p. 6.

[5.15] D. A. Aaker, *Strategic Market Management*, John Wiley & Sons, 1984, p. 161.

[5.16] AAMI Standard, *Standard for Interchangeability and Per-*

formance of Resistive Bridge Type Blood Pressure Transducers, Arlington, VA, USA, 1986.

[5.17] A. Rastegar, New generation of smart sensor signal conditioning, *Proceedings of Sensors Expo 94*, Cleveland, OH, 20–22 September, 1994.

[5.18] J. J. Wortman, J. R. Hauser and R. M. Burger, Effect of mechanical stress on P-N junction device characteristics, *J. Appl. Physics*, 35, No 7, July 1964.

[5.19] W. Włodarski, The possibility of applying the zener diode as a transducer for measuring dynamic pressure, *IEEE Transactions on Industrial Electronics and Control Instrumentation*, May 1972, No. 2.

[5.20] A. Jayaraman, M. E. Sikorski, I. G. Irwin and G. H. Yates, Effect of hydrostatic pressure on P-N junction characteristics and the pressure variation of the band gap, *J. Appl. Phys.* 38 (1967) No 11.

[5.21] K. Petersen, F. Pourahmadi, J. Brown, P. Parsons, M. Skinner and J. Tudor, Resonant beam pressure sensor fabricated with silicon fusion bonding, *Digest of Technical Papers of 1991 International Conference on Solid-State Sensors and Actuators, Transducers'91*, San Francisco, CA, USA, 24–28 June 1991.

6
SENSOR AND CIRCUIT INTEGRATION IN BIOMEDICAL APPLICATIONS

6.1 Introduction

As solid-state integrated sensors have been reduced in size, have become more reliable, and have come to incorporate on-chip signal processing, their use as research tools as well as clinical instruments in biomedical applications has been increasing. Solid-state sensors and actuators find a wide range of applications in biological environments [6.1–6.2]. They are often used for invasive monitoring of patients ranging from simple measurement of electrical and chemical/biological events, such as the recording of biopotential discharges generated electrochemically within individual neurons, to recording and monitoring of physical events. Solid-state actuators are also finding increasing application for delivering electrical and chemical stimuli to the body [6.2]. Eventually, it is hoped that a combination of sensors and actuators will be used to realize complete closed-loop control prosthetic systems that enhance or restore functions for the handicapped, as shown in Fig. 6.1. In such systems, the sensors derive control signals from the body and use these signals to generate appropriate stimuli to induce a desired function in the body. The application of the stimuli is adaptively controlled

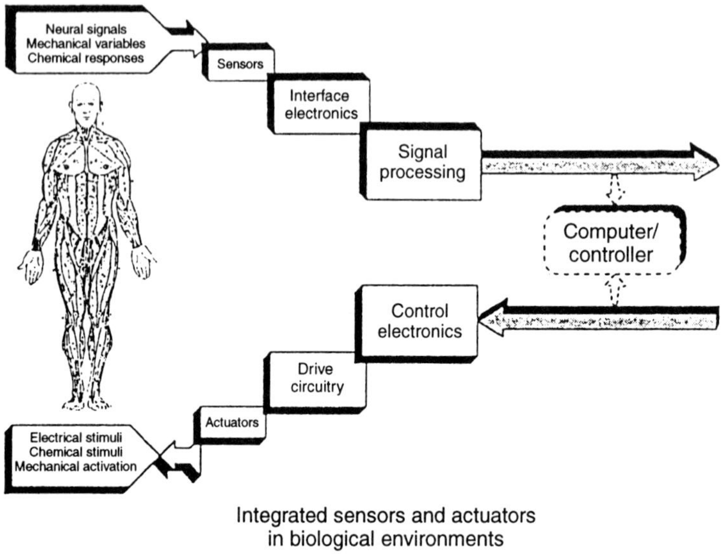

Figure 6.1. *Diagram showing the circuit blocks needed in implantable biomedical systems.*

by the feedback signals provided by the sensors. Once developed, these systems will find widespread application in auditory, visual, and muscular prostheses to restore function to the handicapped [6.3–6.4].

Because of the complexity of the tasks required to implement a successful prosthesis, and since these prostheses have to operate in the harsh environment of the body for extended periods of time, their reliability and stability has to be ensured before they are committed to actual use. It is these two characteristics, in addition to others such as sensitivity and resolution, that make solid-state transducers particularly suitable and indeed required in biomedical applications. Achieving these two specific requirements is to some degree possible by the integration of electronics along with the sensors and actuators that are implanted in the body. This chapter discusses the challenges and requirements for the successful on-chip integration of electronics with solid-state biomedical transducers.

Since the details of the actual integration of various circuit technologies with solid-state transducers are discussed in chapter 1, this chapter concentrates on the most important requirements and challenges that face investigators developing integrated biomedical sensors and actuators. First, a review of typical applications and transducers used in biomedical applications is provided. This review identifies the types of devices needed and the characteristics they have to provide. Next, the challenges and requirements for these transducers are discussed in detail, focusing on the two main areas of circuit and technology requirements. This is followed by a discussion of possible approaches to the achievement of these requirements. Next, examples of a few implantable biomedical systems are presented. Finally, some conclusions are drawn and views on future trends and remaining challenges are presented.

6.2 Applications

Applications for implantable biomedical systems are too numerous to discuss in detail here. Generally, two categories for biomedical applications of transducers can be identified: (1) monitoring and measurement of various physical and chemical parameters and (2) delivery of some stimulus or signal to the body in order to create a controlled change or function in some part of the body.

The first category is rather broad and covers many specific applications. For example, patient monitoring during both surgery and normal daily life is required to ensure that the parameters of interest are in the range they need to be and that the overall health of the patient is stable. These parameters include blood pressure and pulse rate, and chemical species such as glucose and pH level [6.5]. Another example is the monitoring of electrical biopotentials generated by neurons in order to study and understand the operation of the human nervous system and the organization of complex neural structures. Many of these applications do not necessarily utilize the measurement results to initiate an immediate action, as is necessary in prostheses. That is, the eventual goal is just the measurement of one or more parameters from the body. Examples of this include measurement of pressure, level of activity, temperature, pulse rate, etc. Some of these can be performed using external devices, and some require the implantation of a device into the body. For all of these applications, however, one of the most important requirements is that the overall system be as small as possible so that it does not disturb the activity of the patient, unless the individual is restrained to a bed, as in surgery.

In the second category, the main function is the delivery of a stimulus, either chemical, electrical, or mechanical, to a particular part of the body. In order to achieve this, an actuator of some type is required that can deliver the required stimulus. The delivery is typically performed in conjunction with signals provided by a sensor that is either implanted into the body or resides external to the body. An example in this category is the delivery of electrical stimuli to the cochlea in the human ear in order to produce a sensation of sound in a completely or partially deaf individual [6.4]. These stimuli are in the form of current pulses that are delivered through multielectrode probes that are implanted in the scala tympani. Chemical agents such as insulin can also be delivered for those who suffer from diabetes. In this application, the implantable system should be capable of continually monitoring the blood sugar level, to compare it to a prespecified level stored electronically in the implant, and deliver a suitable amount of insulin from a local reservoir through microvalves incorporated into the implant. In any of these applications the delivery of the required stimulus is initiated by a control signal that is derived either from the body, such as in closed-loop neural prostheses, or from the external environment, as in sounds that are recorded externally and converted into appropriate electrical pulses for the auditory prosthesis mentioned above.

The above two categories are very broad and many other examples can be cited beyond those already presented. Both of the categories, however, require the utilization of systems that can produce the required function, whether sensing or actuation, in a stable and reliable manner. Indeed many of emerging biomedical implants and devices have to go through long and complex trial studies and tests before they are approved for human use not only to show that they can produce the functions they are designed for, but mainly to ensure that their use does not irreversibly harm the individual. An excellent example of a potentially useful but yet impractical sensor for biological applications is the ion-sensitive field-effect transistor (ISFET) [6.5–6.6] used for pH measurement. Although this sensor was developed many years ago, it has been plagued by stability and reliability problems and has not yet found its way into practical biomedical applications. Therefore, it is worth noting again that the principal requirements for most implantable biomedical systems are reliability and stability. In order to achieve these functions one has to design systems that offer certain features. It is in the implementation of these features where solid-state integrated biomedical sensors and actuators offer their most important advantage, and where circuit integration becomes critical. In the next section some of these features and their impact on system reliability are reviewed.

6.3 Challenges and requirements

As mentioned above, in order to provide improved reliability and stability, the biomedical system of interest has to possess a number of features. Since we are primarily discussing the issues regarding the need for circuit integration in biomedical transducers, we will not discuss the reliability challenges posed by the transducer itself. Rather, we will concentrate on the features that circuit integration offers to satisfy the requirements for biomedical systems. These requirements can generally be divided into two categories: (1) the requirements for the electronic circuits in terms of their function and operation and (2) the requirements for circuit and packaging fabrication technologies in implementing the entire system. We will discuss these in more detail below. We will also consider some of the sensor reliability issues pertaining to the fabrication technologies employed where appropriate.

Electronic circuits are pervasive in our daily lives and implement a variety of functions in almost every system that we come in contact with these days. They are becoming more important in biomedical devices as new technologies are developed and as new improved systems are designed. Here, too, they perform many functions. However, for many implantable biomedical systems, one can identify a few very important features that are certainly *needed* and *critical*, while there are others that are *desirable* and *useful*. Implementation of biomedical systems for many applications demands the achievement of three main requirements. These are size reduction, reduction of the number of output leads through multiplexing, and signal amplification and buffering for improving the signal-to-noise ratio of recorded signals. Other desirable functions include self-testing, signal processing and data reduction, programmability, telemetry, and sensor data interpretation. Each of these areas is discussed in more detail below.

6.3.1 Size reduction

The size of any implantable biomedical system has a significant effect on the overall performance, stability, and acceptance of the device in any biological tissue. In both sensing and actuation applications, the device should be small enough so that it can approach the tissue of interest with minimal damage and couple to the tissue as closely as possible. The lesser the damage to the tissue, the better is the quality of the signals that are recorded from the tissue (in the case of implanted sensors), and the more effective is the impact of the stimuli that are delivered to the tissue. In addition, size plays a very important role in biocompatibility which has a central role in the eventual utilization of any implantable device. The body either totally rejects foreign bodies by encapsulating them with fibrous tissue and by generating chemicals that tend to attack them, or reacts violently through infections that eventually damage the area that is under investigation. Reducing the size of an implant can only improve the stability and reliability of an implant.

For some applications, such as in recording of neural signals from the central nervous system, no size is too small, and the system ideally has to have no physical dimensions to it. Obviously this is not possible for implantable systems. However, being usually below a minimum size, which differs from application to application, the implanted device is practically transparent to the tissue and can reside within it without any noticeable effect. For example, when recording from the central nervous system, if we can reduce the size of the silicon carrier that supports the recording sites to below about 30 μm we can effectively make the device transparent to the tissue and minimize disturbance. This is because the dimensions of the biological structures within the cortical tissue (such as the neurons which are about 30–50 μm in diameter, and blood vessels which are larger than $\sim$ 50 μm in diameter) are generally larger than 30 μm and any implant smaller than this will cause minimal perturbation to the surrounding tissue. It is due to the requirement for a very small size, and because of the ever increasing demand on system functionality, that in many implantable systems the electronic circuitry has to be integrated with the transducers that interface with the tissue of interest. Although in many research and clinical applications the circuitry can be hybrid attached to the transducer(s), circuit integration inevitably leads to smaller size and improved reliability. Examples of these will be given later in this chapter.

It should also be noted that one may argue that the inclusion of circuitry on an otherwise passive device (with no circuitry) also leads to an increase in size. Although this is correct in general, we should keep in mind that for many present and future biomedical systems there will be some electronic function that needs to be performed as close to the transducer as possible, which in turn demands that this electronic circuitry be integrated with the transducers. In summary, many next generation biomedical systems will contain a substantial amount of electronic circuitry which should be integrated monolithically with the sensors and actuators in order to reduce the overall size of the system.

In addition to size, circuitry is often needed in biomedical devices for multiplexing and amplification. The signals recorded by many biomedical sensors have to be sent out to the outside world for further processing and storage. These signals are typically very small and the sensor interface typically has a high impedance. In addition, for many implantable transducers it is desirable to record or deliver many signals using a single packaged device. Therefore, the number of leads that need to be attached to the device can become large. Inclusion of interface electronics and signal processing circuitry at the sensor site can help to alleviate these problems and serves two main functions: (1) lead reduction, (2) signal amplification and output lead buffering.

6.3.2 Multiplexing for lead reduction

Reducing the number of output leads is another important function in a majority of biomedical systems. External leads are not only a major source of failure but are also relatively large compared with the device itself. Data multiplexing can not only reduce the amount of circuitry required for the sensor (circuit blocks such as the ADC can be shared among several sensors), but it also reduces the number of external leads by either multiplexing several sensed signals onto a single output lead, or by superimposing clock and control signals over the power lines [6.7]. Low-noise analog multiplexing is an especially important function in sensing arrays and systems which require the simultaneous measurement of many signals in order to accurately measure the parameter of interest. In addition, reducing the number of leads that are interconnected to a sensor package is very important in simplifying the packaging, reducing the cost, and improving the long-term reliability of the system in which the sensor is integrated. In addition to sensing applications, data multiplexing can significantly reduce the number of leads required in an actuating system. In many such applications, often several actuators and delivery channels are incorporated into a single implantable package. Multiplexing allows several control signals, along with clock and power, to be delivered sequentially into the implant and latched for later delivery into the tissue according to a prespecified protocol.

It should be noted that while multiplexing goes a long way in reducing the number of leads, it does not reduce it zero. It is of course desirable and often required that there be no leads at all. This can be done through telemetry of both signals and power to the implant [6.8]. Indeed, implantable telemetry systems are becoming more important, and feasible as new technologies are developed to practically incorporate them into the overall system. The incorporation of telemetry systems into implantable biomedical systems can significantly enhance the overall reliability of the system and will simplify the packaging. It is anticipated that before the end of the century totally implantable telemetry-operated prosthetic systems will be developed for both patient monitoring and prostheses. In fact, most auditory implants in patients today utilize some form of telemetry for either data or power transfer across the skin. An important requirement for the development of a telemetry system is low power dissipation. The microelectronics industry is intensively pursuing the development of a variety of low-power techniques for next generation systems. The results of this development effort will indeed be directly applicable to the design of wireless implantable systems.

6.3.3 Amplification and buffering

The signals recorded by many sensors are typically low in amplitude, and the sensor interface usually exhibits a high impedance at the frequencies of interest. Integration of interface electronics and signal processing circuitry at the sensor site serves a number of functions, including signal amplification, impedance transformation, and signal filtering and buffering.

Amplification of recorded biological signals remains one of the most important functions in many applications where the amplitude of the signal is typically low. Amplification of these signals at the sensor site before transmission to the outside world not only enhances the overall signal-to-noise ratio, thus reducing the effects of noise from the environment, but also allows for the full utilization of the dynamic range of an analog-to-digital converter for those sensors that incorporate an ADC in the sensor module. For many integrated sensors this amplification can be achieved by using MOS and bipolar amplifiers that require only a minimal amount of circuitry with nominal gain, bandwidth, and performance specifications. CMOS amplifiers are perhaps the most suitable since they provide high gain and high input impedance through a relatively simple and compact circuit, and are readily compatible with integration of high-density digital circuitry on the same chip [6.9]. CMOS amplifiers are now easily capable of providing open-loop gains of 90 dB, are fast, can be compensated for their offset using various techniques involving charge storage and cancellation, and consume little power and die area. These amplifiers are 3–5 times smaller than bipolar amplifiers, making it possible to include tens of per-channel amplifiers on a single chip. In addition, the input impedance of CMOS amplifiers is very high, which makes them ideal for sensor applications. Although MOS amplifiers have typically had larger noise levels than their bipolar counterparts, various circuit and fabrication techniques are used to minimize the noise in applications where the noise level of the interface electronics determines the resolution of the sensing system [6.9]. More sophisticated amplifiers incorporate programmable gain capability to maximize signal-to-noise ratio and optimize the amplifier output range for ADCs.

Since many implantable sensors have characteristic impedances that are very high (in the mega-ohm range), it is required that the output leads are buffered to protect against leakage and parasitics in the face of long-term exposure to body fluids. Providing this enhanced drive capability for the output leads is crucial to the long-term stability of the overall implanted system and simplifies the packaging and encapsulation constraints on the output leads. Often the amplification and impedance transformation can be achieved simultaneously using the same amplifier. For some applications where gain is not immediately necessary on the sensor chip, a simple buffer circuit can be integrated in the form of source followers that require only a few transistors and consume very little area and power. For capacitive-based sensors with voltage readout, one has to ensure that the input capacitance of the buffer is as small as possible to minimize capacitive loading. Signal filtering can also be incorporated in amplifiers. For multiplexed multisensor systems, filtering of the signal is often required to prevent such problems as aliasing that can introduce high-frequency noise into the signal passband, thus overriding the low-amplitude sensed signal. Signal filtering can also improve the signal-to-noise ratio of the sensor as it filters the out-of-band high-frequency noise that may be introduced by the electronic devices themselves. A variety of low-pass, bandpass, and high-pass filter circuits are available using switched-capacitor circuit techniques or more conventional resistor-capacitor techniques.

6.3.4 Other functions

In addition to these primary functions, the electronic circuitry is increasingly used to perform functions such as self-testing of the transducers and the circuit blocks, analog-to-digital conversion, offset cancellation to allow full utilization of the dynamic range of the ADC, radio frequency telemetry for wireless communication and power transfer, and on-chip data processing and interpretation. Reducing the data processing overhead on the external computers by signal processing at the sensor site will be required for many future implantable sensors. Biological events have a very low bandwidth and it is often not necessary to transmit the full passband to the outside world. For many applications having knowledge of the occurrence of an event or of the frequency of the occurrence of an event is all that is needed. In these application it is very desirable to perform event detection and discrimination at the sensor site. For example, in closed-loop neural prostheses where the control signals needed for operating the prosthetic device are derived from the neural signals recorded from the nervous system, the useful information is often included in the rate of neural discharges rather than the absolute amplitude of each discharge [6.10]. Therefore, one can use an on-chip processor that transmits only this information and discards the rest. This can effectively increase the bandwidth of the transmission link and can reduce the overhead of the external signal processing units. Since most of the data in this form can be transmitted digitally, the external transmission protocol can be designed around a bus-organized system, thus greatly increasing the data transfer speed and improving the overall signal-to-noise ratio. It is believed that for many future closed-loop prosthetic systems such signal processing has to be performed at each sensor site.

It should again be emphasized that signal amplification, filtering, and multiplexing of several data channels onto a single data line also reduce packaging problems and inherently improve the reliability of the device by enhancing the signal-to-noise ratio and by eliminating the need for too many external leads.

6.4 Technology needs

In order to implement many of the circuit functions discussed above, and in order to minimize size, the circuitry has to be monolithically integrated with the transducers. As a result, suitable fabrication technologies need to be utilized to achieve both of these goals. Other chapters in this book discuss the details of the fabrication techniques for both sensors and circuits in general. Here, we will focus on the technology needs and requirements for both the transducers and the electronics in biomedical applications. The choice of the fabrication technologies for these two areas should be made so that a number of requirements are satisfied.

These requirements include: (1) biocompatibility, (2) appropriate packaging, (3) electrical isolation, (4) thermal isolation and (5) low cost and overall simplicity. These are discussed in more detail below.

6.4.1 Biocompatibility

Any technology used for the fabrication of the transducers, circuits, and the package should be biocompatible with biological tissue and should be able to tolerate exposure to the harsh environment in the body. The issue of biocompatibility cannot be emphasized enough since it is the only requirement that is unique to this application. Because of biocompatibility issues, it is often not possible to use established materials and technologies. For example, many hermetic packaging technologies that are routinely used in the IC industry cannot be used in biomedical devices because of the reaction of biological tissue to them [6.11]. It should be noted that sometimes the circuitry may not come in direct contact with the tissue because it is housed in a small hermetic package of its own. However, sometimes it is separated from the tissue only by a thin film that is an integral part of the circuit process, which has to be biocompatible. In summary, in the area of technology and materials, we need to develop new techniques and materials for the transducers, the circuitry, and the package.

6.4.2 Packaging

Encapsulation and packaging of implantable biomedical systems for long-term implantation in the body is of paramount importance as most of these systems will fail due to attack by biological fluids. Any package used in these systems should not only protect the transducers and their associated circuitry from the body fluids, but it should also allow for contact between the sensing/actuating sites and the tissue. The requirement for this selective packaging has posed many challenges for implantable systems. Circuit integration can often simplify the packaging of biomedical systems. This may seem unjustified at first because adding circuitry to any transducer demands that this circuitry be protected from the outside environment. While this is true, it should be noted that the addition of the circuitry often solves more important problems and challenges. As was mentioned in the previous section, two of the important circuit functions are multiplexing for lead reduction and buffering for added drive capability and lower impedances. Both of these functions significantly simplify the packaging of implantable systems. Leads are often the most damage prone portion of any implantable biomedical device. It is therefore important that we not only reduce their numbers in order to simplify the packaging task and reduce any adverse mechanical trauma that too many leads can induce in the tissue, but also ensure that they are sufficiently buffered so that the electrical system remains operational in case the insulating materials forming the package are partially attacked by biological fluids. Therefore, adding electronics to implantable sensors improves both the mechanical and the electrical performance of the system.

Obviously, the circuitry itself needs to be protected from the biological environment. The degree of difficulty with which the on-chip circuitry can be encapsulated depends strongly on the structure of the transducers and their fabrication process. Hermetic as well as non-hermetic techniques have been and are being developed for encapsulating a variety of these systems. Inorganic thin films such as silicon nitride and silicon dioxide, diamond films, and organic thin film polymers such as polyimides and Parylene C have been used to encapsulate both sensors and electronics [6.12–6.13]. Hermetic packaging using silicon-glass electrostatic bonding and silicon-silicon bonding is becoming more feasible as these technologies mature and as new techniques for low-temperature bonding of glass to silicon are being developed [6.14–6.17]. In summary, the protection of on-chip circuitry for most integrated sensors is not an impossible problem, and very often the advantages of having on-chip circuitry with regard to packaging outweighs the problems and disadvantages.

6.4.3 Electrical isolation

As mentioned above, any biomedical implant should not only be protected from the harsh body fluids, but it should also be designed so that it is does not damage the surrounding tissue. One aspect of this is the proper electrical isolation of parts of the implant that come in contact with the tissue. For many implants, as in recording and stimulating microprobes to be discussed later, usually the silicon substrate that supports the transducers and the electronics will inevitably come in direct contact with the tissue. In these instances, the electrical bias on the silicon substrate that is required to ensure correct circuit operation should be compatible with the body. This means that the substrate should be at ground. This in turn means that the circuit fabrication process should be chosen such that it allows the substrate to be grounded while allowing the other internal junctions in the circuitry be reverse biased for proper circuit operation. For example, a circuit process such as p-MOS which forces the silicon substrate to be an n-type semiconductor that is biased at the supply voltage is not compatible with biomedical applications. While it is possible to encapsulate further this part of the substrate with another insulator, the cost and complexity associated with this additional packaging requirement are prohibitive and it is advantageous to change the circuit technology such that electrical isolation is inherently provided. This will be discussed in more detail when examples of silicon microprobes are presented.

6.4.4 Thermal isolation

An important aspect in the design and packaging of implantable biomedical sensors is that of thermal management and isolation. Any implant should be designed so that its operation will not excessively raise the temperature of the surrounding tissue, otherwise permanent damage will be done. This is particularly important for implants that incorporate internal circuitry. The power dissipated by the circuitry generates heat which should somehow be removed safely by the body. This requires that low-power circuits be employed wherever possible, which narrows the choice of different circuit technologies. This is one of the reasons why low-power CMOS is attractive for biomedical applications. In addition to the choice of circuit technology, the fabrication techniques should be capable of providing layers that can couple to the tissue uniformly and appropriately to remove the generated heat in a consistent manner. In this area the designer may not have much choice, but can design the system so that the hotter portions of the device, where the circuitry is located, are remote from the most sensitive regions of the tissue, where the transducers are located.

6.4.5 Low cost and simplicity

As in any measurement and control system, cost and complexity have to be kept to a minimum when electronics are integrated with transducers. The general requirements for biomedical systems in this area are similar to other application areas. However, as mentioned previously, cost is not always the determining factor in choosing a suitable technology. Rather overall performance, and particularly reliability and stability, are more important for biomedical applications. These performance parameters are often affected by the packaging technique used, and anything that can be done to simplify packaging and improve protection will often lead to lower overall cost. Implanted systems are often required to operate in the body for long periods of time, some for many decades, and it is extremely difficult, costly, and often dangerous, to perform surgery to replace the failed system. Therefore, the determination of cost in these systems should always take these issues into account. It should be noted that, because the inclusion of circuitry in the implant can improve reliability and enables the user to perform tasks that are otherwise not possible, the overall cost of the implant could very well be lower than if one were to consider the cost of silicon only. As mentioned above, reducing size is a major goal in many biomedical applications to reduce damage and improve stability. This will obviously reduce the cost of the device as well. In summary, while total device cost is still important in biomedical applications, it is not the driving factor.

6.5 Approaches

The achievement of the circuit and fabrication needs requires an appropriate choice of overall system partitioning, a suitable utilization of circuit blocks, and development of appropriate fabrication technologies. Discussing all of these in detail is beyond the scope of this chapter, and many of these issues are addressed in other

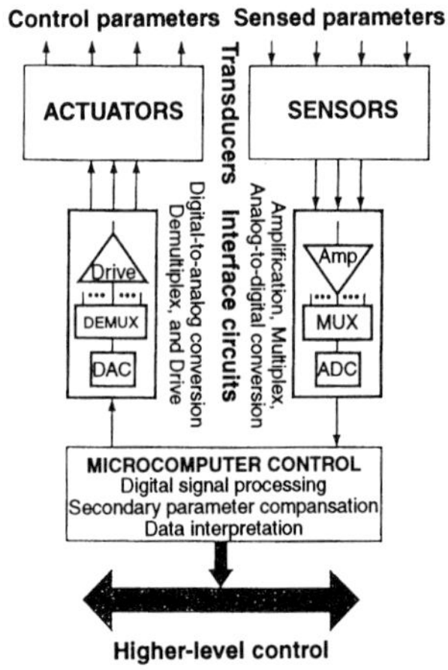

Figure 6.2. *General block diagram of a measurement and control system, showing both the sense and actuate paths between the transducers and a microprocessor control unit [6.1].*

chapters. We will only discuss the issues of relevance to biomedical applications.

6.5.1 *System partitioning of sensor/circuit processes*

Figure 6.2 shows the general block diagram of a measurement and control system, showing both the sense and actuate paths between the transducers and a microprocessor control unit [6.1]. The question of how much circuitry needs to be integrated on a sensor package is one that is very dependent on the specific application. If we assume that there needs to be some circuitry on the same chip as the transducers, then the question becomes more the one of what blocks have to be included in the package. Again, it is incorrect to generalize since biomedical applications vary significantly from area to area. However, based on the discussions in the previous section, there are a few blocks that should certainly be included. These are: the amplifier, multiplexer, and the buffer. The inclusion of these blocks often satisfies most of the important requirements, namely signal enhancement, lead reduction, and output lead drive and buffering. In cases where there is only one transducer on the implant, there is no need for a multiplexer, but the amplifier and buffer function should be included. It can be shown that the integration of these functions on the implant chip will not significantly affect the overall cost of the system compared to the situation when only a few transistors were included [6.18]. This is primarily because these circuit functions can be implemented using only a few hundred transistors and the silicon area required is generally smaller than that required by the transducers themselves. Therefore, even from an economic point of view it is certainly advantageous to add these functions to the transducer(s) chip.

Although it is desirable to include the other blocks such as the circuitry for self-testing and calibration, an analog-to-digital (or digital-to-analog) converter, a microcontroller for data processing, and the appropriate interfaces for communication with a bus, these latter options are primarily decided upon by factors such as performance, cost, and process complexity. In many biomedical applications, often the added value and enhanced performance of adding these functions outweighs the increased cost of adding them to the system. For example, the ability to self-test the implant on demand is a very important function for implantable systems. This is because once these systems are implanted, one does not have access to them for any diagnostic tests unless they are removed, which is extremely costly. Consequently, some minimal self-testing capability should be added whenever possible. Addition of analog-to-digital converters is also becoming increasingly useful and important, especially as wireless communication is becoming more practical and thus required. Wireless transmission of digital information is obviously preferred because it not only improves transmission reliability, but also significantly improves overall bandwidth utilization. Similar arguments can be made for the other circuit functions discussed above. Obviously, biocompatibility should not be impaired by the added functions.

6.5.2 Circuit implementation

The above circuit functions can be realized using a variety of circuit topologies and fabrication technologies. Some of these have been discussed in previous chapters, and we will not present an exhaustive review here. However, it is worthwhile to review possible techniques for implementing the circuit functions discussed above, in particular as far as fabrication of both the transducer and the circuitry is concerned.

The choice of the transducer fabrication technology is often forced by the required performance characteristics. These issues have been discussed in previous chapters and will not be discussed in detail here. The two main sensor fabrication technologies are bulk silicon and surface micromachining, both of which can produce microsensors that are superior in many respects to their discrete counterparts. In the selection of a proper on-chip circuit technology, a number of requirements have to be satisfied. These include high packing density (small die area), low power dissipation (especially for biomedical applications), functional versatility, fabrication simplicity (high yield), and drive capability. Historically, integrated circuits have been fabricated using bipolar and Metal Oxide Semiconductor (MOS) technologies.

Bipolar technologies do not lend themselves easily to the implementation of a number of analog functions that are critical to circuits required in biomedical applications. The first important requirement is the need for high input impedance interfaces, which is usually difficult to achieve with bipolar devices unless area and power-consuming circuit techniques are employed. Secondly, bipolar device structures do not lend themselves easily to analog multiplexing and cannot be used as bilateral switches. Thirdly, and most importantly, low power dissipation is not as easily achieved using bipolar circuits.

MOS circuit technologies, on the other hand, offer very high packing density and low power dissipation, which makes them very suitable for many applications. In addition, their very simple fabrication process results in high yield and increased reliability. They best satisfy the requirements for biomedical sensing systems. MOS devices can be easily used as bilateral analog switches (or transmission gates), which is useful for making a small multiplexer. Although they have a higher 1/f noise than BJT devices, the noise level is still lower than the required level for many applications. The two most frequently used MOS technologies nowadays are Enhancement-Depletion N-channel MOS (E-D NMOS), and CMOS. Because of the above-mentioned characteristics, it is believed that

CMOS is presently best suited as the circuit technology of choice in most biomedical applications. It provides a higher gain per stage than an NMOS circuit, is capable of implementing digital as well as high-performance analog circuits, has a low power dissipation, offers good packing density and high speed, and has a mature fabrication technology that is reliable and gives a high yield.

Although standard CMOS technology satisfies most of the circuit requirements for integrated sensors, the implementation of a number of functions still requires bipolar devices. Some of these functions include voltage references, voltage regulators, low-voltage current sources and sinks, and precision current sources. Voltage regulation is of special interest since as integrated sensors become more autonomous in terms of the amount of circuit functions on them, it will become increasingly desirable for them to handle all their power regulation and referencing locally. This is even more desirable for implantable biomedical sensors that operate using radio frequency telemetry. Based on these overall requirements, a Bipolar-CMOS process technology that offers moderate bipolar device performance in terms of device speed and power handling capabilities will be required for many future integrated sensors and micromechanical systems. The use of parasitic bipolar devices and modified BiCMOS technologies to satisfy this need is perhaps the easiest and least expensive approach [6.19].

Figure 6.3 shows the cross-sectional view of a standard CMOS process that utilizes lateral and parasitic bipolar transistors for circuit applications [6.20]. The CMOS circuitry is integrated on a n-epi layer that is grown over a p-substrate. This layering combination not only permits the fabrication of the CMOS circuitry, but also allows a number of micromachining steps to be performed to fabricate a variety of microstructures. Deep boron diffusion from the front-side of the wafer allows the use of boron etch-stop techniques to be used for forming beams and diaphragms, while electrochemical etch-stops between the n-epi layer and the p-substrate allow the fabrication of a variety of other microstructures, including diaphragms, beams, etc. Ishihara *et al.* have used an electrochemical etching technique to fabricate CMOS circuitry together with a piezoresistive pressure sensor by using electrochemical pn junction etch-stops [6.21]. Although this sensor operates solely using the CMOS circuitry, it is possible to incorporate the parasitic bipolar

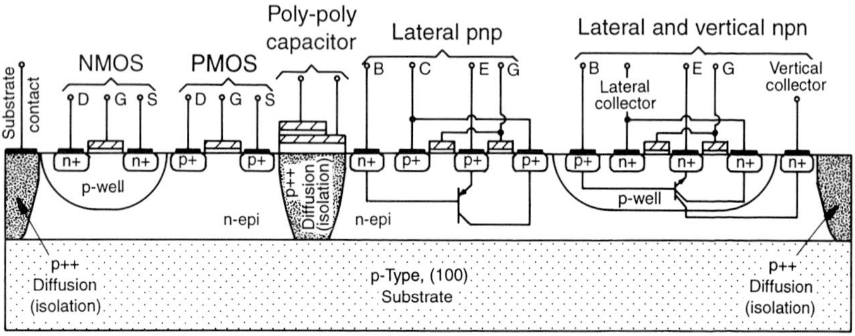

Figure 6.3. *Cross-sectional view of a CMOS-Bipolar micromachining compatible fabrication technology for use in integrated sensors and actuators [6.20].*

devices to implement a number of different circuit functions requiring BJTs.

At the University of Michigan a standard 3 μm bulk p-well CMOS process has been developed together with deep boron diffusion and boron etch-stop to fabricate multichannel active recording electrodes for biomedical applications [6.20]. The process utilizes a total of 10 masking steps, uses two polysilicon layers to form on-chip capacitors needed in many analog circuits, and produces high-performance CMOS devices and circuitry. The details of the process will be presented in the next section where an example of integrated biomedical sensors is presented. In this process, the CMOS circuitry is integrated in the lightly doped n-epi layer grown over a p-substrate. A deep boron diffusion is performed which diffuses through the n-epi layer and connects to the underlying p-substrate, thus creating isolated n-epi regions that will house the circuitry. These p^+-diffused regions will protect against lateral etching of the sensor during the EDP etching process, while the etch from the back-side is timed to prevent the removal of the p-substrate. They are also used to form silicon microstructures for a variety of applications through the use of boron etch-stop techniques in concentration-dependent etchants such as EDP and KOH. In addition, the p^+-diffused regions can be used to provide junction isolation for the CMOS portion of the circuitry and to create isolated n-epi regions that form the collector of high-performance vertical npn BJTs, as shown in Fig. 6.3. Vertical npn bipolar devices with a current gain in the range 100–150 have been obtained using the above process [6.20]. In addition to the vertical transistors, one can utilize lateral npn and pnp bipolar transistors that are naturally formed through the CMOS process [6.19].

Note that in this process two high temperature steps are required, one for the formation of the deep p-wells (this is done at 1200°C for about 16 hours) which will house the n-channel transistors, and the other to form the highly concentrated p^{++} regions (this is done at 1175°C for about 15 hours) that form the silicon microstructures for the transducers. These two high-temperature steps cannot be performed separately because each will significantly affect the other. A modified process has been developed that allows the formation of both the p-well and the p^{++} regions while minimally disturbing the standard CMOS process [6.22]. In this modified process the deep boron diffusion and the p-well formation are combined so that they share the same long-time, high-temperature step, as illustrated in Fig. 6.4. The process starts with the p-well ion implantation, and then a thick thermal silicon dioxide is grown as the masking layer for the deep boron diffusion. After the oxide layer is patterned to define the probe shape, the deep boron diffusion is performed for 16 hours at 1175°C using solid boron oxide sources. During the deep boron diffusion, the implanted ions are driven in to form the p-well. Since the deep boron diffusion is the same as before, it is not affected by the combination. For the p-well, the ion implantation parameters are modified so that after the deep boron diffusion step, the p-well is close to that in the standard CMOS process in terms of the surface impurity concentration and the junction depth. After this step is performed, the wafers are processed with the rest of the standard CMOS steps until the end, when the wafers are encapsulated and then released in a suitable silicon etchant.

Compatible integration of the sensor and the circuitry is perhaps one of the main concerns when fabricating sensors with on-chip circuitry [6.18]. The combined sensor/circuit process should be

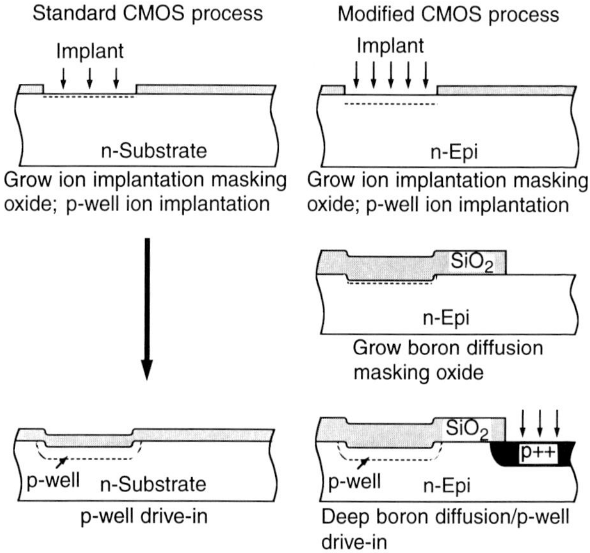

Figure 6.4. *Comparison of the p-well formation in the standard CMOS process and the combined deep boron diffusion/p-well formation in the probe process [6.22].*

such that the circuit process is minimally interrupted by the sensor process, while still allowing precise micromachining of the silicon substrate. As mentioned in the first chapter the ideal merging of the sensor and the circuit processing is best achieved when the sensor process is performed at the beginning or the end of the circuit process. In the sensor-CMOS-bipolar process described above, etch-stops are formed at the beginning of the process in order to avoid exposure of the circuit to high thermal budgets required for the deep boron diffusion, followed by a nearly unaltered circuit process, in this case CMOS. The final silicon etch completes the process. With such a sequence, minimal change is required in the normal circuit recipe. In the case of a deep boron sensor diffusion (for a diaphragm or other microstructures), the p-well drive-in can be performed simultaneously with the boron etch-stop diffusion. For sensors fabricated using surface micromachining technologies, it is also possible to incorporate electronics with the sensor. A number of groups around the world have demonstrated sensor-circuit integration using a standard CMOS process [6.22–6.23].

6.5.3 Packaging technologies

As mentioned in previous sections, the packaging of implantable biomedical systems is a very important issue and challenge that needs to be addressed at a very early stage in the development process. Many of these requirements and challenges are similar to those of other application areas. However, in biomedical applications there are a few unique requirements that pose more limitations on the available choices. Of these, the two most important are: (1) small size and the ability to encapsulate the on-chip circuitry at the chip level while providing selective access to the environment for the transducers and (2) biocompatibility.

Because the size of implantable systems has to be kept to a minimum, traditional packaging techniques used for transducers cannot be used. For example, large hermetic and non-hermetic packages such as plastic or metal enclosures used for IC applications significantly increase the size of the overall implant and are impractical. Many of the materials such as polymers, solders, and glasses cannot be used because of their incompatibility with biological tissues. In addition, since biomedical systems have to operate in harsh biological environments for long periods of time, they should be capable of protecting both the sensors and the circuits using non-traditional materials and packages.

In order to package the integrated transducer chip, most new technologies have focused on the use of a variety of thin inorganic and organic films. These include deposited inorganic thin films of silicon nitride, silicon dioxide and metal shields, and organic polymers such as polyimide, parylene and silicones. Inorganic thin films can be deposited using standard IC fabrication equipment, are compatible with solid-state transducer and circuit fabrication, and can be readily patterned at the wafer level before die separation. They generally provide adequate protection against biological fluids, and in the case of silicon nitride provide a nearly hermetic package for the circuitry. However, they can only be deposited up to thickness of a few microns, and generally require deposition temperatures of several hundred degrees, which limits their use in

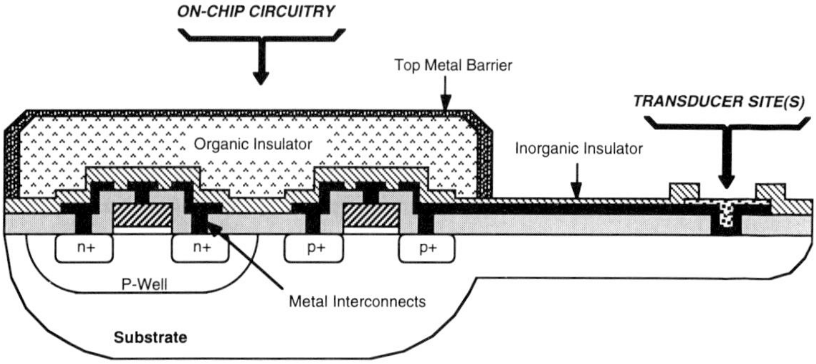

Figure 6.5. *Cross-section of a generic packaging and encapsulation approach for integrated systems used in biomedical applications.*

encapsulating metal interconnects on IC chips. Organic materials can be deposited to thicknesses of several tens of microns, have been shown to provide long-term protection under electrical stress [6.12], and many are biocompatible [6.11]. However, they cannot be easily patterned using photolithographic techniques, and cannot withstand some of the wet etchants, such as EDP, used in silicon micromachining. One of the most promising materials is the silicone rubber that has been widely applied in biomedical applications. In order to take full advantage of the properties of both of these categories of materials, the best choice of packaging technology for biomedical systems seems to be the use of a combination of both of these materials. This is illustrated in Fig. 6.5, where the cross-section of a chip-level encapsulated integrated system is shown.

Several different approaches for encapsulating the various portions of the system is shown here. The circuitry, CMOS in this example, should be completely protected against exposure to the outside environment by insulating the metal interconnect lines. This can be accomplished by first insulating these lines using a deposited inorganic films. These films provide the first level of encapsulation and also are preferred because they provide a suitable surface for the top organic layers to which to bond. LPCVD films of silicon nitride and silicon dioxide are most suitable because they are pinhole free, are dense, and have excellent step coverage as they go over the metal lines [6.24]. However, since they have to be deposited at temperatures of higher than 600°C, they are not compatible with aluminum metallization. In this case, refractory metals and metal silicides such as tungsten and titanium silicides can be used [6.22–6.23]. Several investigators have used these films with some limited success. The high deposition temperatures cause significant mechanical stressing in the films due to the different thermal expansion coefficients between the metals and the deposited dielectrics. Therefore, care has to be taken in using LPCVD films as insulators in order to achieve reproducibility and low contact resistance [6.22–6.23].

It is possible to use low-temperature films of PECVD nitride and low-temperature oxide (LTO). However, in the case of PECVD films, it has been shown that good step coverage cannot be easily achieved, which then results in exposed metal surfaces. For example, Fig. 6.6a shows the cross-section of an aluminum line encapsulated with PECVD oxide and nitride after etching in EDP [6.25]. Note that because of the poor step coverage the aluminum line has been attacked by EDP. Low-temperature pyrolytic oxide films (LTO) have been used successfully and provide excellent step coverage, as shown in Fig. 6.6b. These films, however, cannot provide adequate insulation when used alone and need to be coated with other protective layers. The advantage in using deposited inorganic films is that they can insulate the metal interconnect lines between the circuitry and the transducers, as shown in Fig. 6.5, and can be easily patterned to provide access for the transducer to the outside environment [6.24].

When LTO is used over the circuit regions, it is possible to coat these areas with additional layers of an organic polymer to provide further protection [6.12, 6.25]. For example, polyimides and/or silicone films can be deposited and later patterned, or deposited by casting through a stencil mask [6.26–6.27]. The advantage of these organic films is that they can be deposited in thicker layers and reduce the total electric field drop across the film. This reduced electrical stress significantly improves the lifetime of the package. Since some of the organic films have a tendency to absorb water [6.28], it is possible to deposit a top barrier metal layer over the organic insulator to shield it from direct exposure to body fluids. Metal shields such as gold and tantalum can be deposited and are

(a)

(b)

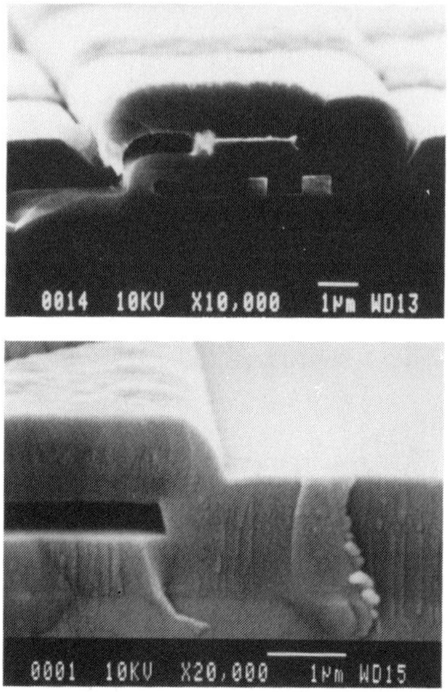

Figure 6.6. *(a) SEM cross-sectional view of an aluminum line encapsulated with PECVD oxide and nitride after etching in EDP [6.25]. Note that because of the poor step coverage the aluminum line has been attacked by EDP. (b) SEM cross-sectional view of an aluminum line encapsulated by low-temperature pyrolytic oxide (LTO) films. Note the excellent step coverage of LTO over the aluminum line.*

inert in the body. This triple-level packaging scheme is very effective and satisfies most of the requirements of biomedical systems. It is small, can be photolithographically patterned, utilizes the advantages of both inorganic and organic films, and is compatible with most technologies.

Encapsulating the transducer poses different challenges since selective access has to be provided to the external environment. The transition region between the insulating film and the transducer opening has to be stable and should not degrade over time. In particular, the insulator has to provide a stable seal to the transducer material in order to prevent any fluids from penetrating underneath and causing delamination of the insulator from the underlying material. Inorganic films are very attractive because they generally adhere well to the underlying material, whether metal or semiconductor [6.25]. Organic films can also be used but require much more care in ensuring that they can properly seal to the underlying films. Consequently, it is preferred that these transducer regions be protected using deposited inorganic films which can be precisely patterned.

As mentioned previously, biocompatibility is a very important criterion in choosing the materials used for packaging implantable systems. Organic films such as silicon dioxide, silicon nitride, and polysilicon have been shown to be biocompatible [6.29–6.30] and are therefore preferred. Inorganic materials such as polyimide, silicone, and parylene are also biocompatible and are preferred for biomedical applications [6.11].

Another technique for encapsulating biomedical devices at the chip level utilizes bonding of two or more chips to provide a hermetic seal that can protect the electronics. Silicon-glass electrostatic bonding [6.14], silicon-silicon eutectic bonding [6.16], or silicon-silicon electrostatic bonding using an intermediate sputtered glass layer [6.15–6.16] have all been used and have been shown to satisfy the needs of their particular application areas. The advantage of many of these packaging techniques is that they can be performed at the wafer level, and can be formed selectively over the electronics. Recent attempts have used custom-designed glass packages for hermetic packaging of the electronics while providing sealed feedthroughs for access to the electronics, and allow access between the sensor and the environment of interest [6.17].

Packaging of solid-state biomedical systems will remain one of the most challenging problems facing the biomedical community. However, it should be noted that by integrating on-chip circuitry it is possible to amplify and buffer the recorded signals before transmission to the outside world and relieve the package from having to protect otherwise weak signals. This is even more important for long-term reliability of the implant. For multisensor devices, multiplexing of several data channels onto a single data line not only relieves packaging and encapsulation but also improves the lifetime and yield of the final product by minimizing the number of leads and bonds, which are potential points of failure. Therefore, the designer should weigh the advantage of improved performance and reduced lead count when incorporating on-chip circuitry against the disadvantage of having to protect this circuitry against harsh environments encountered by the implant. It is believed that for many applications the advantages of incorporating signal processing circuitry on the transducer chip site outweigh the disadvantages.

6.6 Examples

In order to illustrate the issues discussed above, in this section we will present an example of an implantable biomedical system, namely silicon-based implantable multichannel microprobes for recording and stimulation of the central nervous system. These microprobes have been under development at the University of Michigan for the past ten years, and represent one of the most advanced implantable systems under development.

An important area in biomedical applications is that of neurophysiology and the understanding of the operation of the central nervous system. The brain is the control center for all of our bodily functions and transmits the necessary information to various organs for implementing these functions. If we can understand the operation of the complex structure of the central nervous system, if we can receive sensory information from the brain by recording its electrical activity, and if we can deliver control information to it through appropriate electrodes, we can then begin to develop totally closed-loop implantable systems for many prosthetic applications. One of the most widely used techniques for studying the central nervous system at the cellular level is the recording of extracelluar biopotentials generated electrochemically within individual nerve cells. Although the recording of single, isolated neural discharges has been performed for many years, very little has been done to simultaneously record biopotentials from several units. Consequently, while much has been learned about the operation of single neurons, very little is understood about their interaction as functional circuits.

In order to achieve such understanding, the simultaneous recording of discharges from units which are functionally related but physically separated is required [6.10]. The simultaneous recording from many points is also essential to gaining an understanding of neural physiology from general cortical organization, cellular interaction, electrode-cell coupling, and sensing areas into the lesser-known organization of complex systems of neurons. The recording and processing of control signals from many points is also required for the successful development of closed-loop neural prosthesis [6.31]. Most prosthetic devices depend on neuro-electronic recording and stimulation to drive nerve or muscle fibers which are no longer supplied with inputs from their normal sensors or control channels. These devices are currently operated open loop and are therefore relatively limited in their capability. It is clear that useful prostheses will eventually require control signals which are derived from within the body, and should be capable of delivering the proper stimuli to the biological area of interest using precisely delivered and controlled charge packets through multichannel stimulating electrodes. Examples include the auditory prosthesis for the deaf [6.32–6.33], devices to combat incontinence [6.3], and mobility aids that control artificial limbs and reactivate limbs paralyzed by spinal cord injuries [6.3–6.4]. In all these applications, multielectrode probes, with on-chip signal processing electronics which are capable of simultaneously recording/stimulating many neurons, are a necessity. These probes will provide for efficient and practical interaction between biological and microcomputer-based control systems.

Integrated circuit and silicon micromachining technologies have been used to produce multichannel recording and stimulating microprobes for use in a variety of electrophysiological applications [6.34–6.35]. The basic structure of a recording/stimulating silicon microprobe is shown in Fig. 6.7. The probe consists of a precisely micromachined silicon substrate which supports an array of thin-film conductors. These conductors interconnect the recording/stimulating sites along the shank of the probe with signal processing circuitry located at the rear of the substrate. These conductors are insulated above and below by deposited dielectrics, with openings defined photolithographically at the sites and the output pads.

For the successful operation of a multielectrode probe over the long term (chronic), signal processing electronics should be directly integrated on the probe substrate. This minimizes the number of output leads, boosts the signal amplitudes and reduces the sensitivity to the effects of leakage from the output leads which

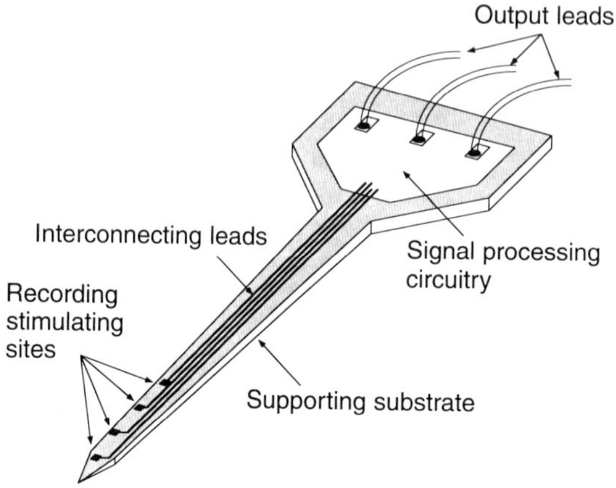

Figure 6.7. *Structure of a solid-state silicon multichannel microprobe [6.7].*

are the most leakage-prone portion of the system. In addition, electronics can be used to change the output data format, provide an on-chip stimulation capability, and allow for remote testing. The inclusion of on-chip circuitry becomes even more important for prosthetic applications where the probes will have to be interfaced to microcomputers for processing of control signals.

6.6.1 Recording microprobes

Figure 6.8 shows the overall system organization of the electronics for an active multichannel recording electrode system [6.36]. The system consists of two circuit portions: (1) implantable electronics for integration directly on the probe substrate and (2) external signal processing electronics for demultiplexing and reconstructing the multiplexed neural signals. The on-chip electronics consists of per-channel preamplifiers, followed by an analog multiplexer and a broadband output buffer to drive the external data line; the circuitry is powered by a regulated battery. The implantable circuitry requires only three output leads for power, ground, and data. In order to allow external regeneration of the on-chip sample clock, a synchronization pulse is inserted once each frame as an

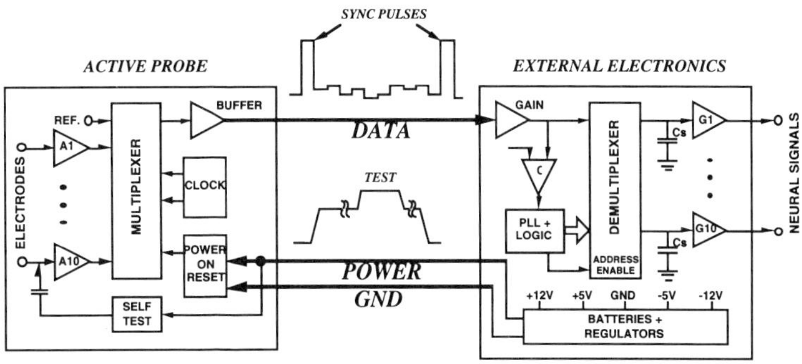

Figure 6.8. *Overall system organization of the electronics for an active multichannel recording electrode system [6.7].*

eleventh channel. The external electronics strips off these pulses, generates the sample clock, and demultiplexes neural signals for external recording/processing equipment [6.7].

Figure 6.9 shows a detailed block diagram of the on-chip implantable electronics for use in a 32-electrode active probe [6.36]. This 32-electrode active probe provides per-channel amplification, analog multiplexing for reducing the output lead count, low output impedance, self-test capability, use of only three output leads, single power supply operation, low power dissipation, electronic positioning and selection of recording sites (by selecting eight of the 32 sites for amplification), DC feedback in the preamplifiers to eliminate the effects of DC potential variations at the recording sites and DC amplifier offsets, and small area.

The on-chip circuitry consists of eight major circuit blocks: the front-end selector, preamplifier, analog multiplexer, clock, self-test, output buffer, I/O control, and input data/clock separator. The 32 electrodes, placed on the micromachined silicon probe shank, are connected to the front-end electrode selector, a 32-to-8 switching array, where eight of the 32 signals are selected. The eight selected signals are then fed into the eight preamplifiers, each having a gain of 150–300 and a 3-dB signal bandwidth from 100 Hz to 10 kHz. The amplified signals are then sampled, multiplexed, and buffered out over the single I/O lead (when the I/O line is set

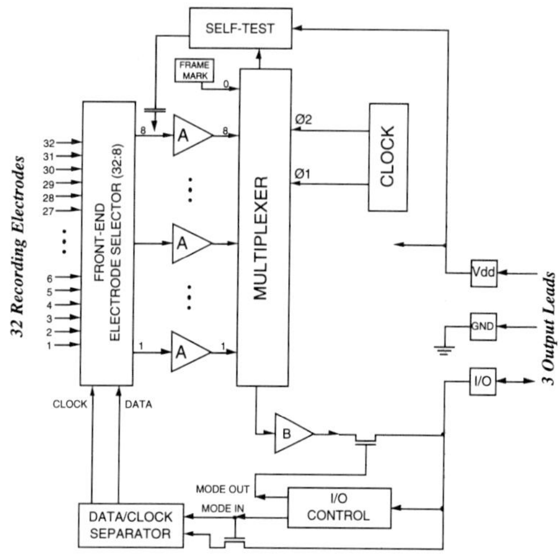

Figure 6.9. *Detailed block diagram of the on-chip implantable electronics for use in a 32-electrode active probe [6.36].*

in the output mode). The multiplexer is controlled by a two-phase non-overlapping clock generated on-chip. The clock signal is sent out together with the multiplexed neural signals in the form of frame marks, which are used by the external electronics to regenerate the clock and reconstruct the analog signals using the external electronics, as discussed above. The external electronics also performs per-channel spike counting and sends the results through the computer interface to the host computer where the results are displayed on the screen. The spike counting results give the user a rough but useful idea of how the selected recording sites couple with active neurons. If the user wants to change the configuration of the sites, (s)he can enter the numbers of the desired sites from the keyboard, and then the computer sends that information to the external electronics. In turn, the external electronics first sends a signal to switch the on-chip I/O controller to the input mode and then superimposes the control data from the computer onto the synchronization clock and sends them over the same I/O line to the probe on-chip data/clock separator.

The data/clock separator separates the control data and the synchronization clock and sends them to the front-end electrode selector. When all the input control data have been shifted into the front-end selector, the I/O line is switched back to the output mode and recording resumes with a new set of recording sites. The I/O line is, therefore, shared by both input and output functions. The power supply line is used not only to supply power but also to control the self-test signal generator.

When testing is desired, the power supply ($V_{\rm DD}$) is pulsed from its normal value of 5 V to 8 V. This latches an on-chip test-enable latch in the TEST mode. After the circuit is latched in this mode, the supply is returned to its normal value. The test-enable latch then turns on the test-waveform generation circuitry. This circuitry receives a 16 kHz frame signal from the shift register and divides it down in frequency to 1 kHz through a 4-stage binary counter. The resultant 5 V, 1 kHz signal is then capacitively attenuated to an amplitude of 50 mV and is then applied capacitively to all the electrodes using 60 fF on-chip coupling capacitors. The amplitudes of the resulting induced electrode voltages are then amplified and multiplexed out in the normal way, providing an indication of the electrode impedance levels and a test of the functionality of the on-chip electronics. After testing, the circuitry can be returned back to its normal mode by turning the power supply off and then back on again.

The front-end electrode selector selects eight electrodes from the 32 that are available and connects them to the eight preamplifiers. The use of such a selector is a tradeoff; many recording sites are required for neurophysiological studies, but not so many can be accommodated with limited power dissipation and chip area. Also, since not all the sites couple to active neurons all the time, with a front-end electrode selector, the limited signal processing resources (both on- and off-chip) can be matched to those sites that are close to active neurons. In addition, the selector can be used to move the sites electronically if the relative position of the probe changes in chronic implant situations.

The complexity of the electrode selector depends on the selection flexibility required in a particular application. For full flexibility in selection, where any eight electrodes can be selected out of the 32, the number of different site combinations is 10,518,300. To

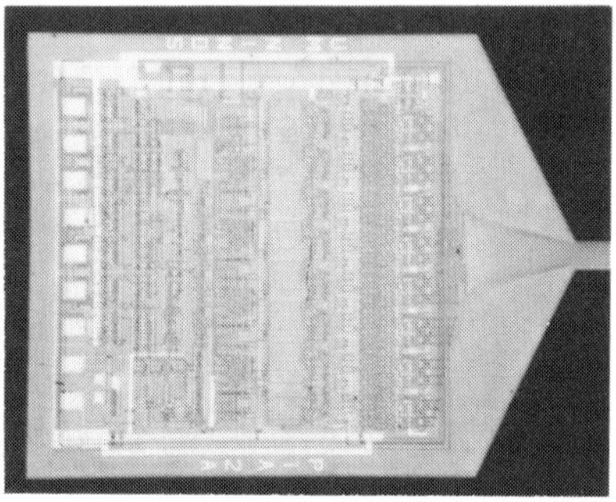

Figure 6.10. *Photomicrograph of a 32-electrode silicon recording active probe [6.36].*

implement such a selector, a very complicated circuit configuration and unallowably large chip area would be required. However, in most neural recording situations, the most useful site configurations would be those that permit depth scanning (i.e. any eight adjacent sites can be selected in depth along the array) and field zooming (i.e. the sites can be changed so that the recording span in depth can be enlarged or reduced). Based on these considerations, a grouping strategy was used to simplify the selector design, and the selector was implemented with eight 4-to-1 multiplexers. In this approach, the 32 recording sites are divided into eight groups, and only one site from each group can be selected at any time. The number of different site combinations possible in this approach is 65,536. This approach still has a large number of site combinations available and allows the most useful configurations such as depth scanning and field zooming to be realized [6.36].

Figure 6.10 shows a photomicrograph of the 32-electrode recording active probe. The circuitry is designed in a p-well 3 μm single-metal double-poly bulk CMOS process, occupies an area of 2.5 mm^2, and dissipates 2.5 mW of power from a single power supply [6.36]. Figure 6.11 shows SEM views of the probe shank that supports the recording electrodes. The probe shank is about 15μm thick, and tapers down to a width of a few microns at the tip. Note that the probe shank is heavily doped with boron and is formed using boron etch-stop technique. This requires that the CMOS circuitry be co-fabricated with the heavily boron doped regions needed

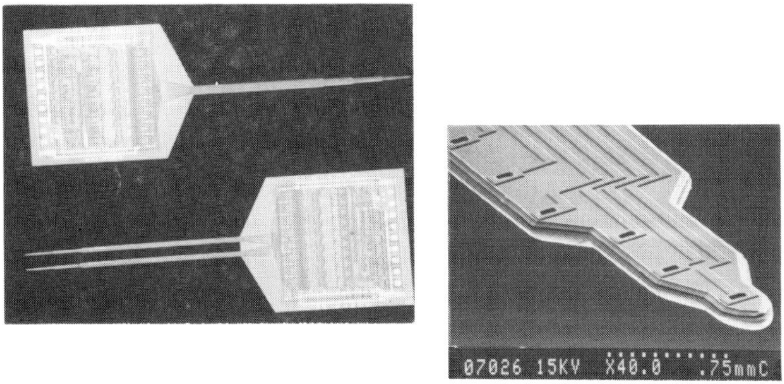

Figure 6.11. *SEM views of the probe shank that supports the recording electrodes. Note that the probe shank is heavily doped with boron and is formed using a boron etch-stop technique. It is about 15 μm thick, and tapers down to a width of a few microns at the tip.*

for sensor fabrication. The combined CMOS silicon micromachining process presented in the previous section is used here, where the p-well for the CMOS circuitry and the p^{++} regions for the probe shank are formed simultaneously through a single high-temperature step at the beginning of the process.

The basic processing steps for the fabrication of the probe and its CMOS circuitry are shown in Fig. 6.12 [6.22]. This p-well CMOS process starts with a silicon wafer of n-type, <100> orientation, 1.0–1.5 Ω cm resistivity and normal thickness. First, a layer of oxide for masking the p-well ion implantation is thermally grown, and patterned to define the p-well regions. The defined p-well regions are then subject to a boron ion implantation, followed by a drive-in step at 1200°C for about 16 hours to form the desired p-well profile. After the p-well is formed, the masking oxide is etched off from the wafer surface, and the LOCOS process begins. A layer of pad oxide is thermally grown, and a layer of LPCVD nitride is deposited. With the active areas protected by photoresist, the nitride is etched off using Reactive Ion Etching (RIE). After a field ion implantation, a thick layer of field oxide is grown around the active regions. The masking nitride and pad oxide are then removed, and the active regions are treated with sacrificial

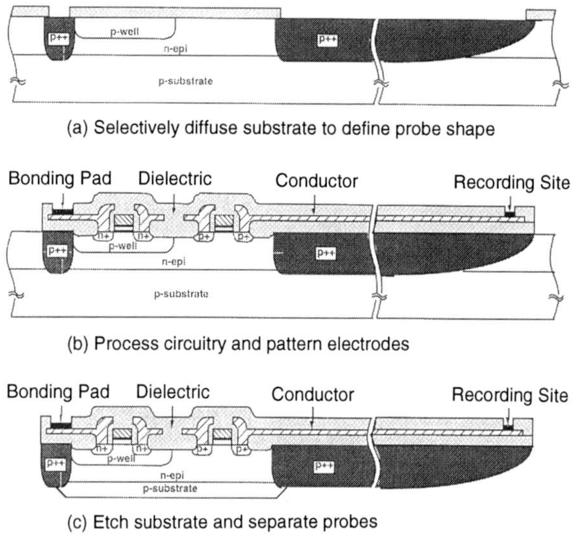

Figure 6.12. *Simplified fabrication sequence for 32-electrode active probe using a modified CMOS process [6.22].*

oxidation, ion implantation for threshold voltage adjustment, gate oxide growth, polysilicon gate deposition and $POCl_3$ doping of the polysilicon. The polysilicon is then patterned with RIE to define the gates and the first level interconnects. Following capacitor oxide growth, a second polysilicon layer is put down and patterned to form capacitors. Then a reoxidation is performed, followed by source/drain ion implantations. A layer of LPCVD isolation oxide is then deposited, and contacts are defined. Finally, aluminum is sputtered, patterned and sintered, and is encapsulated with Low Temperature Oxide (LTO) [6.25]. The field dielectrics are etched from outside of the probe areas, and the wafers are dropped in a solution of EDP for the final wafer dissolution [6.36]. Once the probes become free, and before the p-substrate under the n-epi region is etched, the individual probes are recovered from the etch and rinsed, ready for mounting and assembly.

In the development of this fabrication technology, three major considerations had to be taken into account. First, as mentioned in the previous section, since the probe substrate comes into contact with the biological tissue, it cannot be biased. This means that the substrate has to be grounded, which forces the use of a p-type material. Therefore, in order to utilize a CMOS technology, an n-type epitaxial layer has to be grown over the p-substrate to house both the n-channel transistors (which are located inside a p-well formed in the n-epi layer), and the p-channel transistors. The n-epi can now be biased at the supply voltage to provide junction isolation between it and the p-substrate. Note that the n-epi regions are surrounded on all sides by a p-type material (p^{++} diffused regions around the perimeter, and p-substrate below the n-epi). Second, the metal had to be prevented from being exposed to the EDP solution and the biological fluids. As was discussed before, two approaches were taken; one of these utilized tungsten silicide metal interconnects was encapsulated with LPCVD oxide and nitride layers, and the other utilized aluminum interconnects with LTO on top for protection. The details of these processes are discussed elsewhere [6.25–6.36]. Third, the thicknesses of the dielectric layers on the probe shank had to be chosen carefully so that after probe release the dielectrics would remain in a neutral stress condition. We have used the standard LOCOS oxide and nitride dielectric layers available in the CMOS process. The LOCOS oxide and nitride layers also serve as a boron out-doping barrier on the deep boron doped areas to prevent excessive boron out-diffusion during the long field oxidation step.

This process preserves the ability to define precisely the shape and dimensions of the probe shank, which is critical in reducing damage to the surrounding tissue. In addition, the disturbance to the standard CMOS process is kept to a minimum by adding the micromachining steps at the beginning (when the p^{++} regions are formed), and at the end (when the top insulation is deposited and final EDP etch is performed) of the process. We are routinely fabricating both recording and stimulating probes using this technology, and have applied it to several other non-biomedical sensors [6.37].

The development of implantable multichannel recording systems promises to have a significant impact on many neurophysiological studies and on the development of neural prostheses. In particular, it is expected that these systems will be used in the development of both visual and auditory prostheses. In these applications, however, in addition to multichannel recording probes, one needs to develop multichannel stimulating microprobes that can deliver precise charge packets into the neural tissue for invoking appropriate responses. Multichannel stimulating probes have been implemented using the same technology as recording microprobes and will be discussed in more detail below.

6.6.2 Stimulating microprobes

Electrical stimulation is usually achieved by using microelectrodes to deliver small currents into the tissue for neural activation. Stimulating probes for many prosthetic applications should satisfy a number of requirements. They should be able to deliver current into tissue through multiple sites precisely spaced on a substrate that is small in size to minimize damage to the surrounding tissue. The stimulating sites should be made out of a material that is biocompatible and that does not degrade in terms of its charge delivery capabilities after years of operation in the biological tissue. In addition, the stimulating sites should possess a high charge delivery capacity so that large amounts of charge can be delivered through small sites. One of the most important requirements of a multichannel stimulating electrode is that a minimum number of leads be connected to it. This is required for improved long-term reliability of the implant, for reducing the tethering forces exerted on the implanted probe due to the interconnect leads, and for simplifying packaging and assembly. In order to achieve these requirements active multichannel stimulating probes with on-chip multiplexing and current drivers have been developed [6.35].

The overall structure of active stimulating microprobes is the same as that of the recording microprobes shown in Fig. 6.7. The probe is implemented using silicon micromachining to form probe shanks that support an array of stimulating sites, and houses multiplexing and drive electronics on its back end, fabricated using CMOS technology. The block diagram of a 16-channel active stimulating microprobe is shown in Fig. 6.13 [6.35]. This device has a digital-to-analog-converter (DAC) for every channel, allowing all sixteen channels to be actively stimulating at any given time, with individual control over the current amplitudes. The circuit receives 16-bit data words that are shifted into the data pad using a 4 MHz clock. The timing diagram for the data word is also shown in Fig.

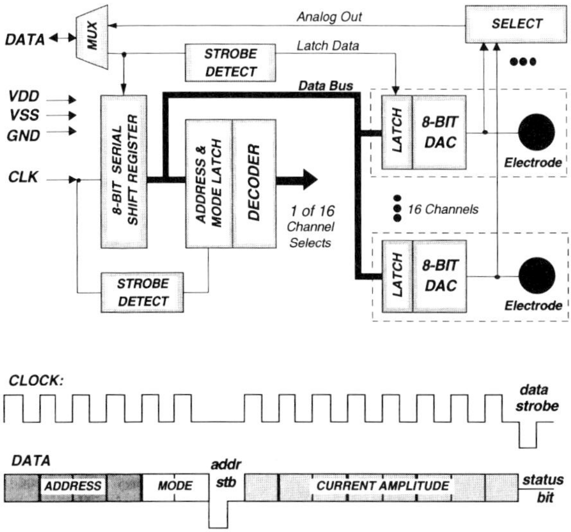

Figure 6.13. *Block diagram of a 16-channel active stimulating microprobe [6.35].*

6.13. Four bits represent the site address, two are mode bits, eight set the current amplitude, and two are used for strobing and status. In the normal mode of operation, the specified current is set up on the addressed channel and remains on until the channel is accessed again to alter it. Other channels are set up sequentially to source or sink current as desired. At a clock rate of 4 MHz, the time resolution between any two changes in the probe current pattern is 4 μs, which is effectively simultaneous to the tissue. A second mode of operation grounds the associated stimulus channel, allowing the DAC currents to be calibrated externally without stimulating the tissue. A third mode of operation allows the analog voltage developed in response to stimulus current to be observed externally to check site integrity at any time. Finally, a fourth mode of operation biases the electrode above ground between pulses to increase the charge delivery of the iridium oxide electrode sites [6.38].

Other important circuit features of this 16-electrode stimulating probe include analog access through the data pad to all sites in parallel when the circuit is first powered up. This allows the growth of iridium oxide on all of the sites simultaneously. As a safety feature, the probe also includes per-channel timeouts which guard against overstimulation by shutting off the current on any channel that has not been accessed in more than one millisecond. In addition, the last bit in the data input stream is used to allow the probe to signal the outside world if any of the probe input signals are out of their proper range. Thus, probe function is checked with every input frame [6.35].

In addition to the difference in the function of the circuitry between the recording and stimulating probes, the stimulating sites are also made from different materials in order to satisfy the requirements needed in stimulation applications. The stimulating sites are required to meet very heavy demands for chronic stimulation applications. They must be able to withstand long-term exposure to body electrolytes and should be capable of delivering large amounts of charge without dissolution into the body or delamination from the conductor surface. Furthermore, the charge density must not exceed the gas evolution threshold at the stimulating site. A material that has shown promise in meeting the requirements for high charge capacity stimulating sites is iridium. Iridium is the densest periodic element known, it is highly resistive to corrosion under static potential conditions, and is chemically inactive since it is a noble metal [6.35]. This material will grow a hydrous oxide layer on its surface when electrically activated in acidic or saline solutions [6.38]. The oxide exhibits a high charge delivery capability that is believed to be a consequence of oxidation state changes that occur in response to varying voltage levels [6.38]. The deposition of iridium in the stimulating sites is easily incorporated into the CMOS process because it is usually performed after the CMOS processing has been completed. After the deposition of the top dielectric material over the metal interconnect lines, these dielectrics are removed by plasma etching from the stimulating sites and the bonding pads, and iridium is sputtered and patterned using a lift-off process. Therefore, this portion of the sensor fabrication is independent of the CMOS process steps and does not interfere with the standard processing of the wafers.

Figure 6.14 shows a photograph of a fabricated stimulating microprobe. This probe consists of two shanks each of which supports 8 stimulating sites. Each shank is 3 mm long and 135μm

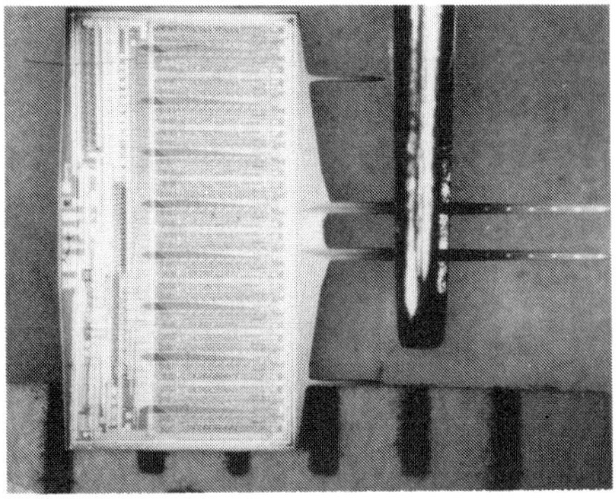

Figure 6.14. *Photograph of a fabricated stimulating microprobe. This probe consists of two shanks, each of which supports 8 stimulating sites [6.35].*

wide, and the shanks are 500μm apart. The site areas are 1000μm^2 and are spaced 250μm apart. The circuitry is fabricated using the 3μm p-well single-metal single-poly bulk CMOS process presented above, occupies an area of 4.4mm $\times$ 2.5mm, contains about 7100 transistors, consumes 80μW of power when idle, operates from $\pm$5 V supplies, and operates at clock rates as high as 10 MHz. The output currents are linear to 8 bits with sourcing and sinking currents matched within $\pm$1/2 LSB, ensuring adequate charge balancing at the stimulating sites [6.35].

The above two examples of implantable recording and stimulating biomedical systems demonstrate the advantages of adding electronics on the same chip as the transducer. As evident, the primary advantages of adding electronics in these two applications are: (1) lead reduction in both devices through the appropriate use of multiplexing schemes, (2) amplification and signal buffering to improve the signal-to-noise ratio and simplify packaging and (3) much added functionality in terms of user control over probe functions and self-testing capability. Future generations of these probes will include low-power analog-to-digital converters, wireless telemetry for power and data transfer, and on-chip signal processing and data interpretation for on-site control and data handling [6.39]. This trend of adding sophistication and versatility to implantable biomedical systems by appropriate use of on-chip electronics will certainly continue in the future.

6.7 Conclusions

The integration of electronics and transducer(s) for the development of implantable biomedical systems has been discussed and reviewed. Because of their difficult requirements and harsh operating environment, implantable systems pose many challenges to the fabrication technologies used for their implementation. These challenges are often simplified by adding on-chip circuitry to an otherwise passive transducer. The ability to amplify low-level biological signals on-chip, to multiplex several signals into a single output lead, thus reducing the total number of wires attached to the implant, and to buffer the output wires is very important to the realizing of reliable and stable implantable systems for a variety of applications. These circuit features often simplify the package requirements and reduce the long-term effects of prolonged exposure to harsh biological fluids. It is the conclusion of this chapter that with today's technology, it is feasible to fabricate a variety of implantable sensors and actuators with on-chip circuitry and improve the overall performance of the system. While there are many factors that influence the partitioning of the system, the choice of the fabrication technology, the design of the particular circuit blocks and the selection of appropriate materials for packaging, the fact remains that for most applications monolithic integration of transducers and circuitry is possible, desirable, and economically feasible. Examples of multichannel recording and stimulating microprobes for use in the central nervous system were given to illustrate the issues and demonstrate our existing capabilities. However, we still have a long way to go before we can implement practical implantable systems for biomedical applications.

Any implantable system for use in human beings has to be reliable, stable, and transparent to its user both esthetically and functionally. All researchers working toward the development of these systems have to illustrate these features in any system before these systems are put to actual use. There is no doubt that circuit incorporation and integration will continue to play an important role in the development of implantable biomedical systems. In fact, in some circles, as in groups involved in the development of closed-loop neural prostheses, it is believed that because of the enormous complexity of biological systems we will never have enough capability in terms of computational power, number of transducers, signal processing capability, small size, and speed, to truly match the performance of biological systems; no matter what we get we will need more. However, it has been demonstrated that we can develop implantable systems that are useful to people. Auditory implants that have helped bring some limited hearing to hearing impaired people are an excellent example of this. These systems have been implanted in patients and are becoming more accepted as our knowledge about their use increases. Many groups are working on the development of visual prostheses [6.40], which is a much more formidable task. We need many more transducers, much faster operation, more sophisticated signal processing and data handling capability, lower power dissipation, and much better materials for packaging and interfacing with the biological tissue.

Consequently, we believe that we will see the continued development of more sophisticated systems based on the full utilization of electronics in implantable systems. Several key areas that will continue to see intense research and development are: (1) the development of low-power circuitry and electronics to prolong battery life and reduce heat generation, both of which are of critical importance in biological applications, (2) the development of small, low-power, long-range wireless telemetry techniques and circuits for power and data transfer to and from implantable systems, (3) the development of new materials and techniques for packaging of implantable systems, (4) the development of more reliable and smaller sensors and actuators for interfacing with a variety of biological tissues, and (5) the development of improved control algorithms and techniques for closed-loop operation of biomedical systems in a number of application areas. Integration of on-chip circuitry along with many transducers will be needed for achieving most of these objectives.

6.8 References

[6.1] K.D. Wise, K. Najafi, VLSI sensors in medicine, in *VLSI in Medicine*, Norman G. Einspruch and Robert D. Gold, eds, Academic Press, 1988.

[6.2] *Special Issue of the IEEE Engineering in Medicine and Biology*, Vol. 13, No. 3, June/July 1994.

[6.3] F.T. Hambrecht and J.B. Reswick (eds) in *Functional Electrical Stimulation: Applications in Neural Prosthesis*, Marcel Dekker, New York, 1977.

[6.4] W. Agnew and D. Mc Creery (eds), *Neural Prosthesis: Fundamental Studies*, Prentice Hall, Englewood Cliffs, NJ, USA, 1990.

[6.5] W. Göpel, New Materials and Transducers for Chemical Sensors, *Sensors and Actuators*, B18–19 (1994) 1–21.

[6.6] H.S. Wong and M.H. White, A self-contained CMOS integrated pH sensor, *Technical Digest, Int. Electron Devices Meeting*, pp. 658–661, San Francisco, CA, December 1988.

[6.7] K. Najafi and K.D. Wise, An Implantable Multielectrode Array with On-Chip Signal Processing, *IEEE J. Solid-State Circuits*, Vol. SC-21, No. 6, pp. 1035–1044, December 1986.

[6.8] T. Akin, B. Ziaie and K. Najafi, RF telemetry powering and control of hermetically sealed integrated sensors and actuators, *Technical Digest*, *IEEE Solid-State Sensor and Actuator Workshop*, pp. 145–148, Hilton-Head, SC, June 1990.

[6.9] P.R. Gray and R.G. Meyer, MOS operational amplifier design; a tutorial overview, *IEEE J. Solid-State Circuits*, Vol. SC-17, pp. 969–983, December 1982.

[6.10] J. Kruger, Simultaneous recordings from many cerebral neurons: techniques and results, *Rev. Physiol. Biochem. Pharmacol.*, Vol. 98, pp. 177–233, 1983.

[6.11] G.E. Loeb, A.E. Walker, S. Uematsu and B.W. Konigsmark, Histologic reaction to various conductive and dielectric films chronically implanted in the subdural space, *J. Biomed. Mater. Res.*, Vol. 11, pp. 195–210, 1977.

[6.12] K. Najafi, J. Ji and K.D. Wise, Multichannel intracortical recording microprobes: scaling limitations, device characteristics, and circuit encapsulation, *Proceedings Transducers '87, International Conference on Solid-State Sensors and Actuators*, Tokyo, Japan, pp. 65–68, June 1987.

[6.13] G.E. Loeb, M.J. Bak, M. Salcman and E.M. Schmidt, Parylene as a chemically stable, reproducible microelectrode insulator, *IEEE Trans. Biomed. Eng.*, Vol. BME-24, No. 2, March 1977.

[6.14] G.D. Wallis and D.I. Pomerantz, Field-assisted glass-metal sealing, *J. Appl. Phys.*, Vol. 40, No. 10, pp. 3946–3949, September 1969.

[6.15] M. Esashi, A. Nakano, S. Shoji and H. Hebiguchi, Low-temperature silicon-to-silicon anodic bonding with intermediate low melting point glass, *Sensors and Actuators*, A21–23 (1990) 931–934.

[6.16] L.A. Field and R.S. Muller, Fusing silicon wafers with low melting temperature glass, *Sensors and Actuators*, A21–23 (1990) 935–938.

[6.17] B. Ziaie, J. Von Arx, M. Nardin and K. Najafi, A hermetic packaging technology with multiple feedthroughs for integrated sensors and actuators, *International Conference on Solid-State Sensors and Actuators*, pp. 266–269, Yokohama, Japan, June 1993.

[6.18] R.F. Wolffenbuttel, Fabrication compatibility of integrated silicon smart physical sensors, *Sensors and Actuators*, A41–42 (1994) 11–28.

[6.19] E.A. Vittoz, MOS transistors operated in the lateral bipolar mode and their applications in CMOS technology, *IEEE J. Solid-State Circuits*, Vol. SC-18, No. 3, pp. 273–279, June 1983.

[6.20] B. Ziaie, Y. Gianchandani and K. Najafi, A high-current IrOx thin-film neuromuscular microstimulator, to be Published, *6th Int. Conf. on Solid-State Sensors and Actuators*, San Francisco, June 1991.

[6.21] T. Ishihara, K. Suzuki, S. Suwazono, M. Hirata and H. Tanigawa, CMOS integrated silicon pressure sensor, *IEEE J. Solid-State Circuits*, Vol. SC-22, pp. 151–156, Apr. 1987.

[6.22] J. Ji and K.D. Wise, An implantable CMOS analog signal processor for multiplexed microelectrode recording arrays, *Digest, IEEE Solid-State Sensor and Actuator Workshop*, pp. 107–110, Hilton-Head, SC, June 1990.

[6.23] W. Yun, R.T. Howe and P.R. Gray, Surface micromachined, digitally force-balanced accelerometer with integrated CMOS detection circuitry, *Digest, IEEE Solid-State Sensor and Actuator Workshop*, p. 126, HiltonHead, SC, June 1992.

[6.24] J.F. Hetke, J.L. Lund, K. Najafi, K.D. Wise and D.J. Anderson, Silicon ribbon cables for chronically-implantable microelectrode arrays, *IEEE Trans. Biomed. Eng.*, Vol. BME-41, No. 4, pp. 314–321, April 1994.

[6.25] J.L. Lund and K.D. Wise, Chip-level encapsulation of implantable CMOS microelectrode arrays, *Digest, IEEE Solid-State Sensor and Actuator Workshop*, pp. 29–32, Hilton-Head, SC, June 1994.

[6.26] H.D. Goldberg, D.P. Liu, R.W. Hower, M.E. Poplawski, and R.B. Brown, Screen Printing: A Technology for Partitioning Integrated Microsensor Processing, *Digest, IEEE Solid-State Sensor and Actuator Workshop*, pp. 140–143, Hilton-Head, SC, June 1992.

[6.27] M.E. Poplawski, R.W. Hower and R.B. Brown, A simple packaging process for chemical sensors, *Digest, IEEE Solid-State Sensor and Actuator Workshop*, pp. 25–28, Hilton-Head, SC, June 1994.

[6.28] M.C. Buncick and D.D. Denton, Effects of aging on polyimide: a study of bulk and surface chemistry, *Digest, IEEE Solid-State Sensor and Actuator Workshop*, pp. 102–105, Hilton-Head, SC, June 1990.

[6.29] J.K. Niparko, R.A. Altschuler, X. Xue, J.A. Wiler and D.J. Anderson, Surgical implantation and biocompatibility of CNS auditory prostheses, *Annals of Otolaryngology*, vol. 98, pp. 965–970, 1989.

[6.30] J. K. Niparko, R. A. Altschuler, D. A. Evans, X. Xue, J. Farraye and D. J. Anderson, Auditory brainstem prosthesis: biocompatability of stimulation, *Otolaryngol. Head Neck Surg.*, vol. 101, pp. 344–352, 1989.

[6.31] F.T. Hambrecht, Natural and artificial sensors in neural prosthesis, *Technical Digest, IEEE Solid-State Sensor Conference*, pp. 5–7, 1984.

[6.32] E. Hochmair, An eight-channel scala tympani electrode for auditory prosthesis, *IEEE Trans. Biomed. Eng.*, Vol. BME-27, pp. 716–722, 1990.

[6.33] G. May, S.A. Shamma and R.L. White, A tantalum-on-sapphire microelectrode array, *IEEE Trans. Electron Devices*, Vol. ED-29, pp. 1932–1939, 1979.

[6.34] K. Najafi, K.D. Wise and T. Mochizuki, A high-yield ic-compatible multichannel recording array, *IEEE Trans. on*

Electron Devices, Vol. ED32, No. 7, pp. 1206–1211, 1985.

[6.35] S.J. Tanghe and K.D. Wise, A 16-channel CMOS neural stimulating array, *IEEE J. Solid-State Circuits*, 27, No. 12, pp. 1819–1825, 1992.

[6.36] J. Ji and K.D. Wise, An implantable CMOS circuit interface for multiplexed microelectrode recording arrays, *IEEE J. of Solid-State Circuits*, Vol. 27, No. 3, pp. 433–443, 1992.

[6.37] E. Yoon and K.D. Wise, A multi-element monolithic mass flowmeter with onchip CMOS readout electronics, *Technical Digest, IEEE Solid-State Sensor and Actuator Workshop*, pp. 161–164, Hilton-Head, SC, June 1990.

[6.38] L.S. Robblee and T.L. Rose, Electrochemical guidelines for selection of protocols and electrode materials for neural stimulation, in *Neural Prosthesis: Fundamental Studies*, W. Agnew and D. McCreery (eds), Prentice Hall, Englewood Cliffs, NJ, ch. 2, 1990.

[6.39] Tayfun Akin and K. Najafi, A telemetrically powered and controlled implantable neural recording system with CMOS interface circuitry, *Proceedings, IEEE 7th Mediterranean Electrotechnical Conference*, Antalya, Turkey, pp. 545–548, April 1994.

[6.40] P.K. Campbell, K.E. Jones, R.J. Huber, K.W. Horch and R.A. Normann, A silicon-based, three-dimensional neural interface: manufacturing processes for an intracortical electrode array, *IEEE Trans. Biomedical Eng.*, Vol. BME-38, pp. 758–767, 1991.

7
SMART SENSORS IN AUTOMOBILES

7.1 Introduction

Sensing elements have been used in automobiles since the late 19th century. For example, the timing of the sparking was achieved via the use of a magneto drive shaft. In the 1920s many countries introduced speed limits and thus the need to be able to measure the speed of the vehicle become clear. For this purpose, a simple rotating magnet, driven through a Bowden cable from the rear of the gear box was chosen. This system remained relatively unchanged for 60 years.

In Japan the motor industry started later than in Europe or the United States. The first automobile company in Japan was the Nissan Motor Company Ltd. which was established on 26th December 1933. Initially the company manufactured models based on European models under licence, and the company has continued to grow to become one of the largest motor manufacturers in the world.

The Japanese automotive market has grown rapidly in recent years and the number of automobiles owned in Japan has reached approximately 65 million, which is almost 10% of the world total. The development of automotive electronics can be divided into 4 phases. The first phase, which took place in the 1950s, was the replacement of some mechanical functions with electronics. The

1970s saw the second phase where mechanics and electronics were truly integrated, and the 1980s saw the development of cooperative and synthetic control systems. The fourth phase, which is now underway, is the development of smart automotive systems, and with this fourth stage has come a rapid increase in the requirement for electronic sensors.

With this increase of the complexity of the automotive system has come the need for greater expenditure on research. To meet this demand Nissan completed, in April 1987, a new clean room facility for the development of electronic and sensor systems. This clean room has a total area of $1500m^2$ and is of class 10. Such developments show the importance attached to electronic systems in automobiles for the 21st century.

There is a wide area of applications for sensors in the modern automobile; some of the more important of these are:

- Ignition control
- Knock sensing
- Fuel control
- Emission control
- In-cylinder combustion
- Speed and torque
- Transmission control
- Suspension control
- Anti-lock
- Anti-spin
- Air-bag
- Fuel quantity
- Fuel flow
- Oil level
- Vehicle navigation
- Headway control

If all these systems are to be included into the automobile, then both cost and size must be reduced to a minimum.

There are several factors which form the driving force to increasing the use of automotive electronics. These are:

- Tighter environmental regulations and safety legislation
- Market expectation for driveability and performance

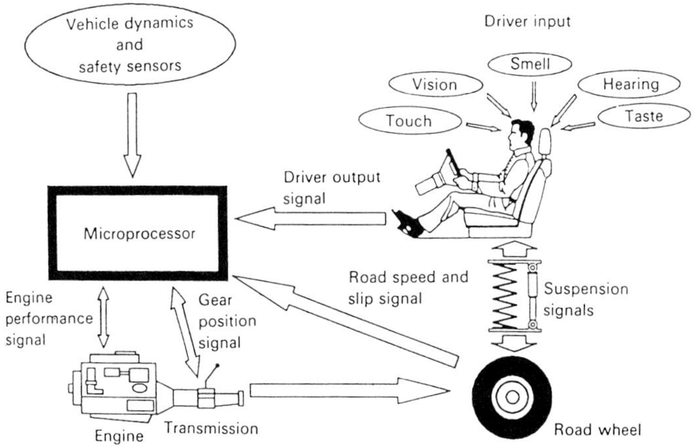

Figure 7.1. *Total sensory system, including driver.*

- Move towards vehicle self-management and diagnostics
- Improved comfort
- Availability of cheap electronics

As these systems develop it should be remembered that, at least for the present, the driver is part of the sensory system and thus all systems should be user-friendly. Such a system is illustrated in Fig. 7.1 [7.1].

When developing sensors for the automobile industry, the system as a whole must be kept in mind. It is possible to choose the optimum sensor for each application and to connect it via a cable to a central control unit. The result of such an approach is shown in Fig. 7.2. and at the other extreme in Fig. 7.3. In this latter case each sensor unit has its own signal processing and is attached to a single bus.

To date, silicon sensors have not seen wide use in automobiles and when they are used they have tended to be in the hybrid form. Whether the sensor be in a smart or hybrid form, the automobile represents an extremely harsh environment. The environmental problems which the silicon device must face are [7.2]:

- Dust, humidity and aggressive fluids

Figure 7.2. *Automobile where wiring has not been optimized.*

Figure 7.3. *Automobile with complete bus system.*

- Ambient temperatures: $-40°C \ldots +85°C \ldots +140°C$
- Shock, vibrations and accelerations
- Disturbances such as transient voltages and electromagnetic radiation

and within this environment the sensor should ideally run for 17 years, or 300,000 km with an operating lifetime of 6,000 hours. This appears to be a very unsuitable environment for silicon sensors. However, not all devices are exposed to every item listed above and there remains a wide range of potential applications for smart silicon sensors. Packaging is also an important consideration in such an environment.

If these problems can be solved there are many advantages with the use of an integrated solution. These are:

- Quality and reliability
- Reduction in cost (potential)
- Reduction in wiring and mechanical components
- Reduction in space requirements
- Improved functionality
- Improved diagnosis

As shown in Fig. 7.2 a simple stand-alone approach to sensor and electronic systems in automobiles limits the range of devices which can be applied. The question is how far to move towards smart sensors, and furthermore what is the definition of a smart sensor. The three options are shown in Fig. 7.4. The first is clearly not a smart sensor option since the sensing element sends a uncompensated analog signal down the cable to the signal processing unit. In some cases this may be the only option since the sensor environment is too harsh for silicon sensors. The second option, Fig. 7.4b, can be described as a smart sensor since the sensor and signal processing are incorporated in a single package. The definition of a smart sensor used in this text is where the sensor and signal processing are included on a single chip, as in Fig. 7.4c.

There has been considerable interest in the application of smart silicon sensors to automobiles and in this chapter the sensor requirement, for each application, is discussed along with the arguments for and against the use of smart sensors.

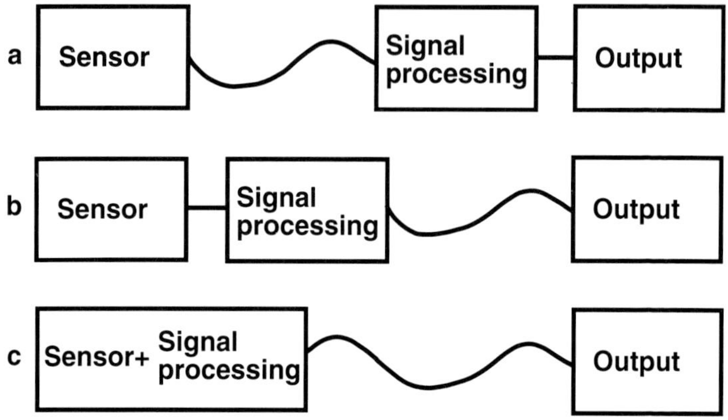

Figure 7.4. *Three basic types of sensor systems; (a) separate sensor and signal processing, (b) hybrid solution with sensor and signal processing in a single package and, (c) smart sensor.*

7.2 Development of automotive electronics

Before the 1970s automobiles contained minimal electronics. In the 1970s the introduction of the microcomputer to automobiles led to a rapid increase in the use of electrical systems for engine control, air conditioning, chassis control, safety and information systems. An overview of the development of electronic systems in this application is shown in Fig. 7.5.

In 1993 the total sales of automotive electronics in Japan was 330 billion Yen (US$3.3 billion) and the distribution of these sales in terms of applications and devices is shown in Fig. 7.6. It can be seen that sensors accounted for 29% of the total sales with the largest portion of sales being attributed to electronic devices such as Engine Control module (ECM) and displays. This figure also shows the distribution of sales in terms of where the electronics are applied. It can be seen that the engine accounts for more that 50% of this. Surveys in Japan indicate that the application of electronics will continue to grow, reaching sales of over US$4 billion by 1998. The growth of the market is shown in Fig. 7.7.

An indication of the importance attached to a device comes from

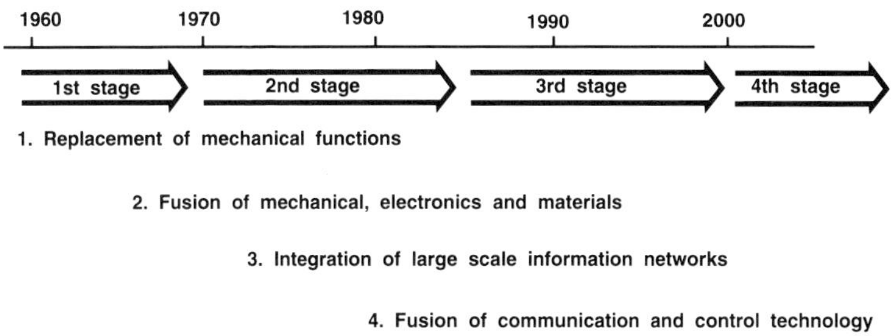

Figure 7.5. *Development of automotive electronics since the 1960s.*

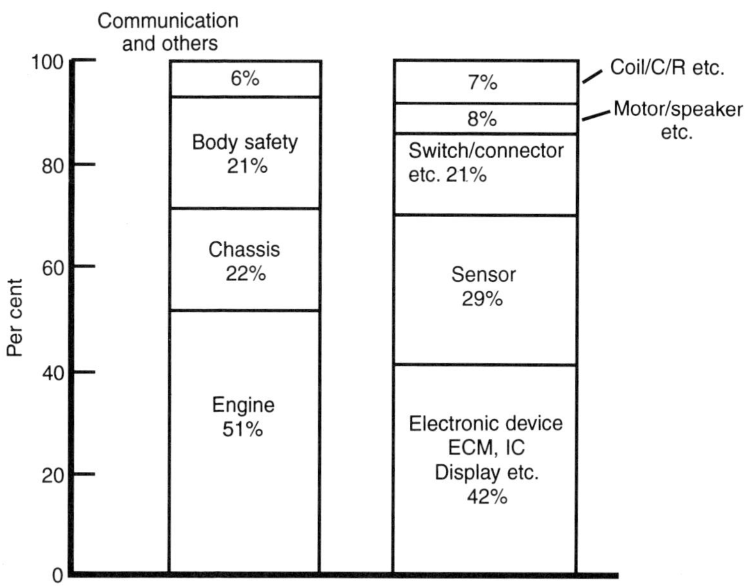

Figure 7.6. *Distribution of automotive electronics in the 1993 total sales of 300 billion Yen. Sources: Report of the Electronic Industries Association of Japan and Report of the Sougou giken Corporation.*

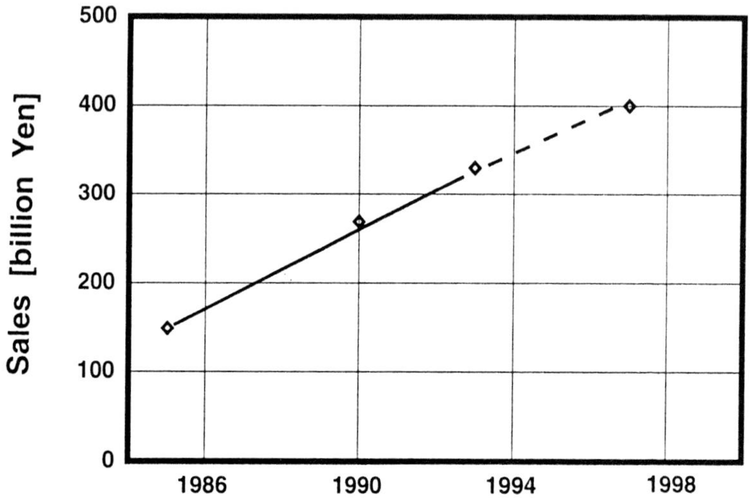

Figure 7.7. *Increase in the Japanese market for automotive electronics. Sources: Report of the Electronic Industries Association of Japan and Report of the Sougou giken Corporation.*

Table 7.1 Patent applications from Nissan and Toyota (1986–1991)

Type of device	% appl.	% increase
Direction, distance, level, position	8	+81
Velocity, acceleration, yaw rate, shock	10	+49
Length, displacement, angle, profile	15	+38
Testing;engine, chassis, structure	11	+11
Force, torque, pressure, power, efficiency	9	+10
Composition, density, viscosity	21	+9

the level of patent applications. Table 7.1 shows that direction, distance, level and position sensors showed the greatest increase, with the applications in the period 1986–1991 being 81% higher than in the previous 6 years. Table 7.2 shows the areas which have seen a fall in interest. These figures do not indicate that the

Table 7.2 Falling patent applications

Type of device	% decrease
Vibration, sound	54
Flow, liquid level	57
Temperature, heat	64

devices are no longer used, but rather that fewer new devices are in development.

7.3 Considerations when applying smart sensors

When applying smart sensors to automobiles both the technical and economic aspects must be considered and these are discussed in more detail below. Smart sensors should be applied when:

1. New smart sensors offer a lower price for the same performance.
2. A higher performance can be achieved. However the price should be kept as low as possible (most customers do not wish to pay higher prices for the new systems).
3. Smart sensors with standard digital I/O signal are attractive for a digital bus in cars, but the cost of implementation will delay their application.

They should *not* be applied when:

1. The performance is lower than that of existing sensors.
2. The electronics would be exposed to electromagnetic radiation or high temperatures.
3. The sensor output signal must be carefully adjusted to the different Electronic Control Unit (ECU) in different cars.

All measurand domains can be addressed using silicon sensors but in each case some problems have be overcome both in technical and financial terms. Examples of each domain are given in Table 7.3 [7.3]. Until recently the considerations listed in this section have prevented the application of smart sensors and in fact silicon sensors were not applied to automobiles in large numbers. The

Table 7.3 Sensors which can be fabricated in silicon by standard IC processes

Sensor type	Main problems
1. Optical	
Photodiode for sensing	Encapsulation
brightness	
color	
radiation	
Charge-coupled devices	Encapsulation
phototransistor	
position-sensitive devices	
2. Mechanical	
Piezoresistors for sensing	Micromachining
force	
pressure	
acceleration	
Capacitive techniques for	Encapsulation
displacement	
tactile sensing	
3. Thermal	
pn junctions for temperature	Self-heating
Thermocouples for temperature and heat flux	Encapsulation
4. Magnetic sensors	
Hall effect for proximity, rotation rate	None
5. Chemical sensors	Special processing
Chemocapacitors for moisture	
ISFET for chem. conc., pH	
SAW (Surface Acoustic Wave) device for gas composition	

development of practical automotive sensors, using the example of a model introduced in 1990, is discussed in the following section.

7.4 Applying sensors in automobiles

As mentioned above, the use of automotive electronics has been increasing steadily since the 1960s. When the Nissan model Cima was updated in 1990 it was fitted with a greatly increased number of electronic components. The automobile is equipped with twenty-

Table 7.4 Types of devices and technologies used

	Ceramics	Metal	Semicond.	Polymer	Total
Position		Mechanical contact lead switch		Conductive plastic	19
Revolution		Induction coil lead switch	Photo transistor, Photo diode		8
Acceleration		Differential transmission Mechanical contact			8
Temperature	Thermistor				9
Flow		Platinum film			1
Level		Lead switch		Conductive plastic	2
Pressure	Piezoelectric device				7
Gas	Zirconia electric cell		Photo transistor, Photo diode		2
Light			Photo transistor, Photocond. device		4
Direction	Fluxgate device (ferrite)				1
Yaw rate			Optical gyro		1
Electricity		Electromagnetic relay	Transistor		2
Total	18	27	8	11	64

three electronic systems (excluding audio systems), sixty-four sensors, of forty-five different types, and ninety-nine actuators. Table 7.4 list the devices used; the majority of silicon sensors in this application are photo-devices while the majority of sensors are metal based. Table 7.5 shows where in the automobile the sensors are applied.

A modern engine management system requires a range of sen-

Table 7.5 Silicon and non-silicon sensors applied in the Cima

Application	Non-silicon sensors	Silicon sensors
Engine	12	1
Chassis	19	0
Air conditioner	11	2
Meter/body	11	2
Safety	3	3

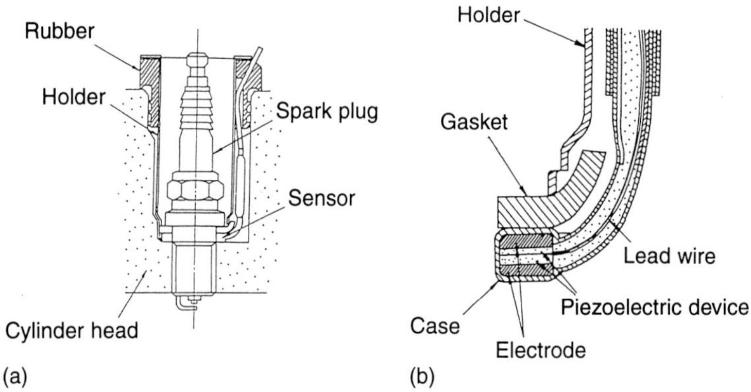

Figure 7.8. *Installation of cylinder pressure sensor introduced by Nissan. (a) Location of the sensor; (b) cross-sectional view of the sensor.*

sors. For example, knocking control requires that the sparking time is controlled; improvement to this increases engine performance. Initially these devices were magneto-based structures but piezoelectric devices, shown in Fig. 7.8, have been applied in Nissan automobiles since 1986. The cost of the piezoelectric device is about US$1, but with packaging, the total cost of the sensor comes to US$10.

In 1992 Toyota introduced a lean-burn system which operated by measuring the internal cylinder pressure. This system, which is shown in Fig. 7.9, measures unstable combustion and is able to increase the air-to-fuel ratio to the optimum value, via a feed-back

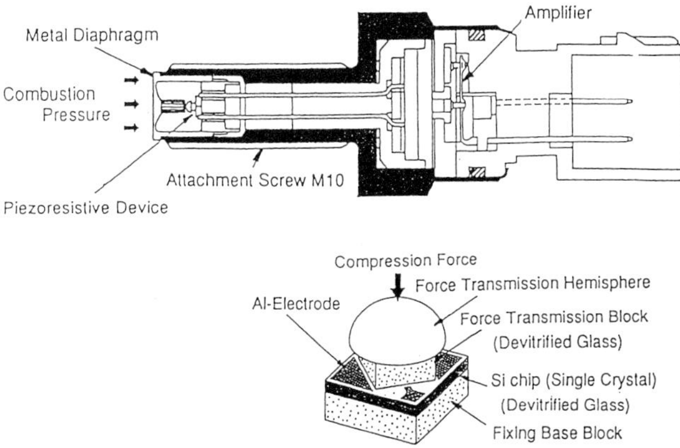

Figure 7.9. *Cross-section of cylinder pressure sensor with piezoresistive device.*

system. This in turn increases the fuel efficiency of the vehicle. The required specifications are:

- Working pressure range — 0–10 MPa
- Burst pressure — 20 MPa
- Temperature range
  - Metal diaphragm — −30–+400°C
  - Force detection device — −30–+150°C
  - Signal conditioning — −30–+120°C
- Sensitivity at 80–120°C — $0.3\ \mathrm{V\ MPa^{-1}}$
- Total allowable error — $0.3\ \mathrm{V\ MPa^{-1}}$
- Measurable frequency — 0.2–5 kHz
- Vibration survival — 30 g
- Power supply voltage — 5 ± 0.25 V

One of the main problems which had to be overcome during the development of this device was keeping the high temperatures away from the electronic devices. This is therefore a case where it is difficult to use a smart sensor.

A hydraulic suspension system is designed to improve driving stability as well as decreasing up/down chassis vibration. The

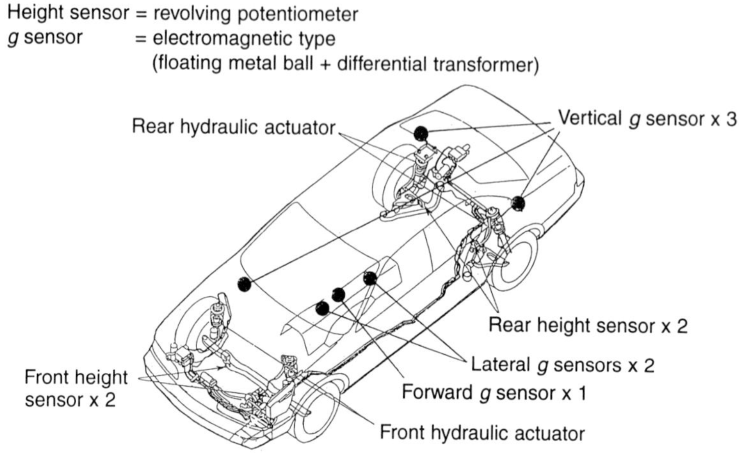

Figure 7.10. *Sensors used in the Nissan active suspension system introduced in 1989. Source: Nissan Technical Review 1989.*

system introduced in Nissan automobiles in 1989 is shown in Fig. 7.10. Although silicon sensors were not used in these models, this system has high smart sensor application potential, which will be discussed below. Accelerometers are one of the most promising areas for smart silicon sensors; silicon accelerometers were introduced in the Toyota Crown in 1992 for the Anti-lock Breaking System (ABS), as illustrated in Fig. 7.11.

A system for controlling the suspension damping force in response to the road conditions was first introduced by Nissan in 1984. This system, which is illustrated in Fig. 7.12, uses an ultrasonic transmitter and receiver to measure the distance between the chassis and road. The returning signal gives information on the condition of the road surface. A similar system, which previews the road ahead of the vehicle by projecting the signal forward, was introduced by Mitsubishi in 1992.

There has been considerable interest in ABS systems in recent years and there has been pressure on the automotive developers to achieve prices low enough so that it can be introduced as standard. A system for measuring wheel rotation for this application was

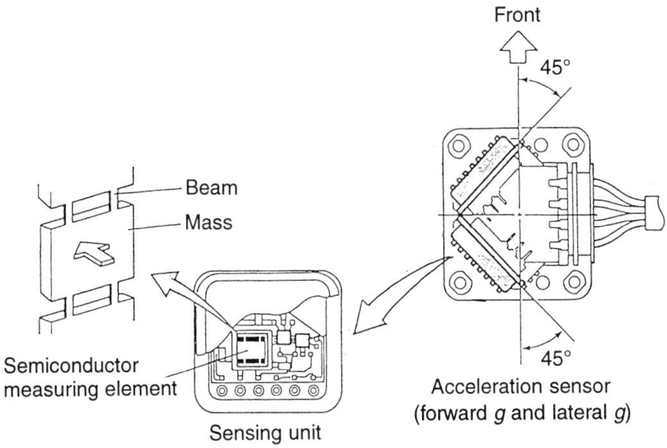

Figure 7.11. *New silicon accelerometer introduced by Toyota in 1992. Source: Toyota Technical Review, vol. 43, no. 1, May 1993.*

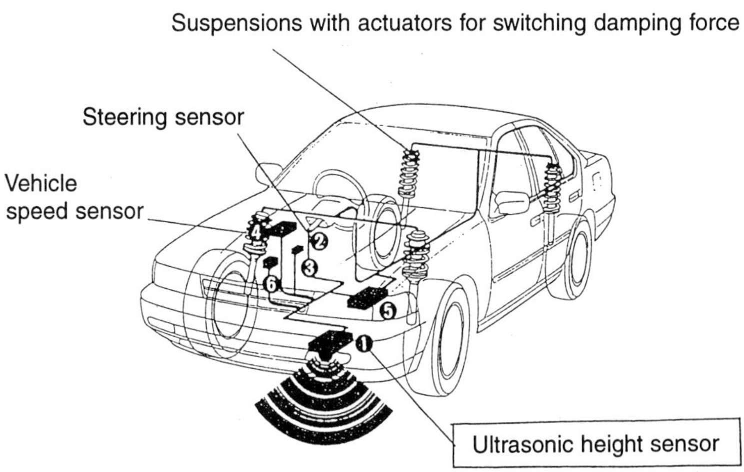

Figure 7.12. *Ultrasonic road surface sensor for suspension control. Source: SAE paper 851652, Electronically controlled shock absorber system used as a road sensor which utilized supersonic waves.*

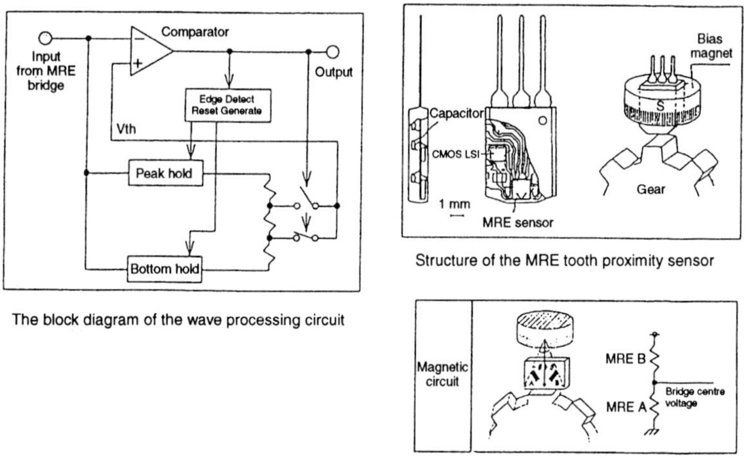

Figure 7.13. *Gear tooth proximity sensor with a ferro-magnetic resistance element. Source: Technical Digest of the 12th Sensor Symposium, 1994, p. 241.*

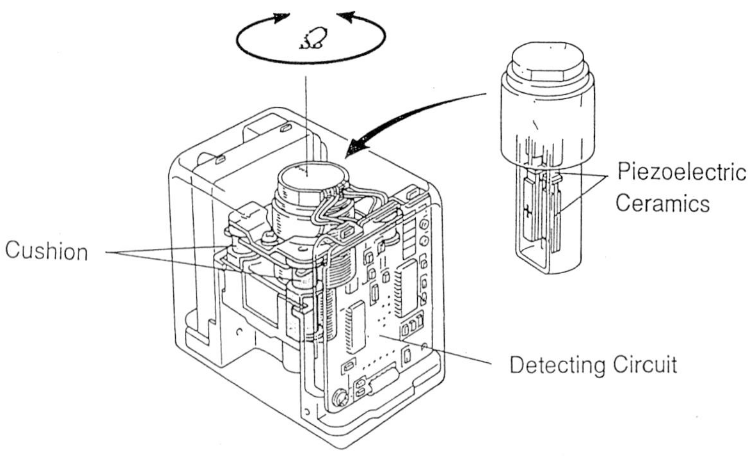

Figure 7.14. *Yaw rate sensor assembly. Source: J. Soc. Automotive Engnrs Japan, vol. 45, 1991.*

marketed by Nippondenso in 1994. This device, based on the magnetoresistors, is at present a hybrid device and is illustrated in Fig. 7.13. A single chip solution would result in a further reduction in size and cost, and this is a likely development for the future.

With the increasing demand for navigation systems has come the need for direction sensors. One such device is the yaw rate sensor which detects the rotation of the vehicle. Such a system, introduced by Toyota in 1991, is shown in Fig. 7.14. However, further reductions in cost are required and present research is directed towards silicon smart sensors. Devices such as micromachined vibration gyros [7.4] are of particular interest for the future.

7.5 Examples of silicon sensors

In this section examples of silicon sensors developed with automotive applications in mind will be discussed.

The application of silicon sensors has been the subject of both fundamental research and development for several years. Optical and magnetic sensors are widely used to detect the engine crank angle and steering angle and this calls for high resolution. A smart optical angle-of-rotation sensor has been developed by Nissan [7.4]. This device is based on the measurement of light passing through a slit in a rotating disc, as shown in Fig. 7.15. Basically the sensor detects the movement of the light edge front as it passes over the chip. This device was later modified to use a simple flip-flop sensor in digital mode to achieve a further reduction in size [7.5]. Each cell gives a signal that its illumination is above a given threshold and thus the output is of a digital nature. This single chip device contains its own voltage stabilizer and oscillator and yields a digital output. A block diagram of the chip is given in Fig. 7.16.

Pressure sensors have been shown to be an extremely large market for silicon devices. These are usually either piezoresistive or capacitive. An example of a smart silicon pressure sensor developed by Nippondenso is given in Figs 7.17 and 7.18 [7.1]. The sensing element is a simple piezoresistive bridge formed on a micromachined diaphragm. On-chip laser trimming of trimming resistors is used to adjust offset, sensitivity and temperature effects. The dimensions and specifications of the chip are given in Tables 7.6 and 7.7.

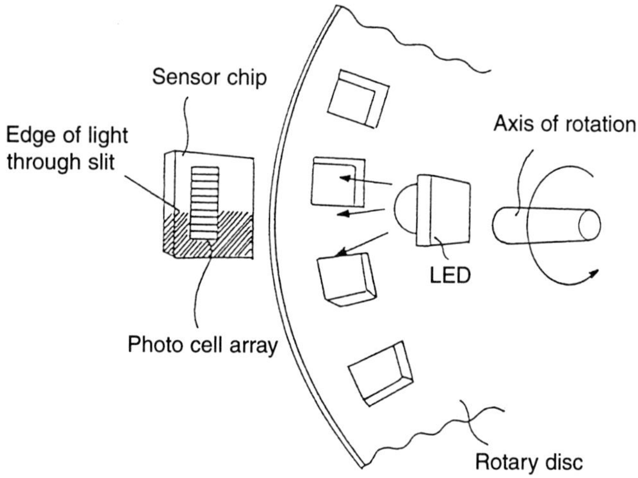

Figure 7.15. *Optical angle-of-rotation sensor setup [7.4–7.5].*

Table 7.6 Chip dimensions

Sensor	{100} plane, p-type gauge/full bridge
Circuit configuration	Op-amp+thin film resistor (bipolar IC)
Diaphragm size	1.8×1.8 mm^2
Diaphragm thickness	35 μm
Chip size	3.5×3.5 mm^2

This sensor makes use of the excellent mechanical properties of silicon. In applications where the environment permits, the smart pressure sensor, it is highly suitable for automotive use. A further development would be such a system with a digital output.

A sensor which has received considerable attention in recent years is the accelerometer. This interest has been mainly sparked by the introduction of the air-bag system, which detects a rapid deceleration, due to a collision, and inflates a bag to protect the driver. The deceleration of interest is usually in the region of 50 g. There are, however, a range of applications for accelerometers,

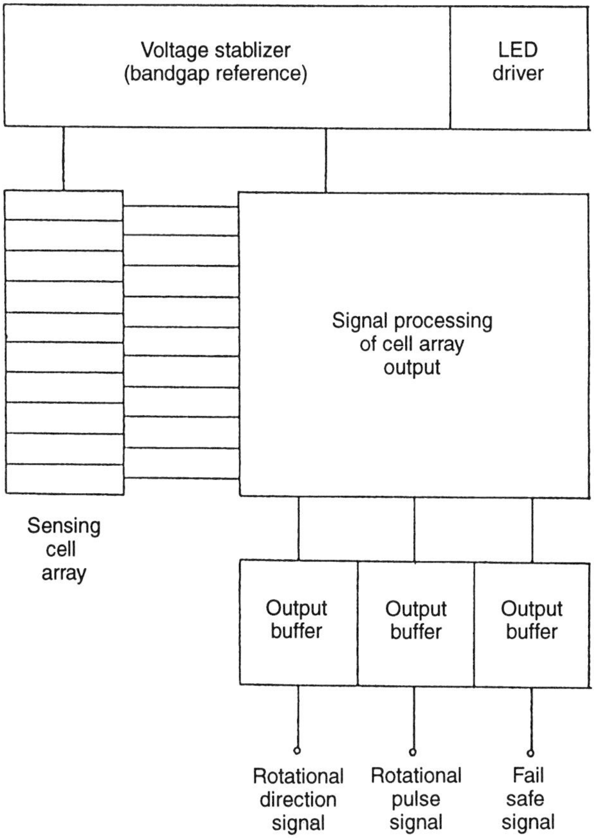

Figure 7.16. *Block diagram of the smart angle-of-rotation sensor [7.4–7.5].*

Table 7.7 Specification of intelligent pressure sensor

Supply voltage	5 ± 0.25 V (DC)
Operating pressure range	150–750 Torr (abs)
Operating temperature range	-30–120°C
Output voltage range	1.2–3.6 V
Accuracy	± 1 Torr

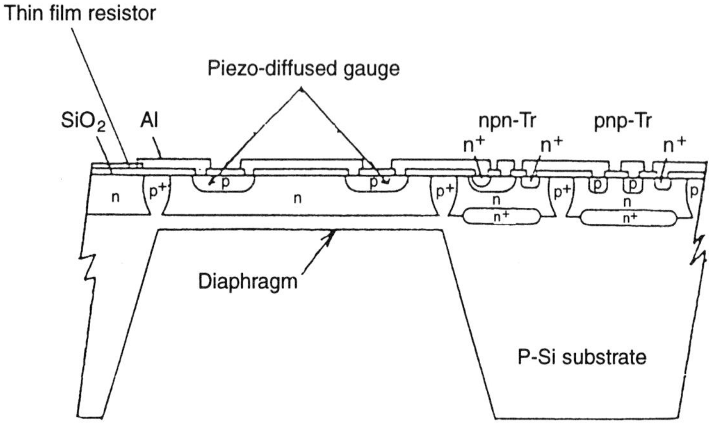

Figure 7.17. *Cross-section of a smart pressure sensor [7.1].*

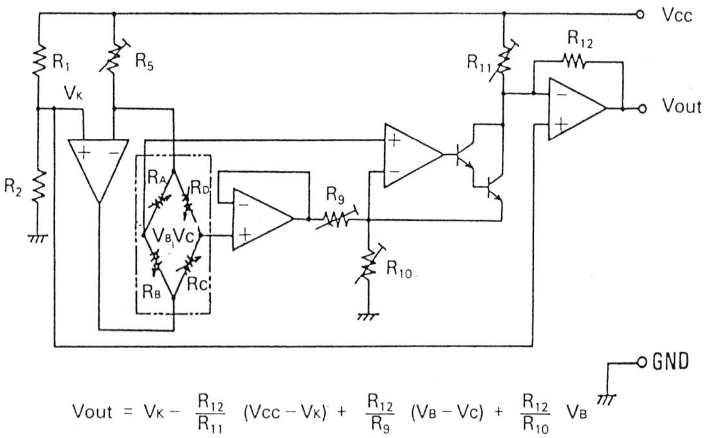

Figure 7.18. *Circuit configuration of a smart pressure sensor [7.1].*

each having its own specific requirements. This are listed in Table 7.8 [7.6].

Although systems such as magnetic-based accelerometers have been proposed, the main contenders are piezoelectric, piezoresistive or capacitive techniques. A comparison of these techniques is listed

Table 7.8 Accelerometer performance requirements for automotive applications

Range	±2 g (vertical body motion) ±1 g (horizontal body motion) ±40 g (wheel motion)
Accuracy	±2% (5% at temp. extremes)
Cross-axis sensitivity	< 1 to 3%
Shock survivability	> 500 g
Frequency range	0.5 to 50 Hz vertical 0–50 Hz horizontally
Temperature range	−40 to +125°C (up to +85°C in cab)

Table 7.9 Comparison of various accelerometer types

Parameter	Piezo-electric	Piezoresistive Silicon	Piezoresistive Thick film	Capacitive
DC response	No	Yes	Yes	Yes
Bandwidth	Wide	Moderate	Low	Wide
Self-generating	Yes	No	No	No
Impedance	High	Low	Low	Very high
Signal level	High	Low	Low	Moderate
Temperature range	−55–100	−55–150	−50–120	−200–200
Linearity	Good	Moderate	Moderate	Excellent
Static calibration	No	Yes	Yes	Yes
Cost	High	Low	Low	High
Ruggedness	Good	Moderate	Moderate	Good
Withstands shock	Yes	No	No	No

in Table 7.9. From this table it would appear that only piezoelectric accelerometers are suitable for automotive applications since the other options are less rugged and not able to sustain a shock. However, silicon piezoresistive and capacitive accelerometers have been applied to automobiles. This has been achieved by using suitable overload protection and damping. Some of the delay in incorporating these sensors has been to ensure that the device met the reliability requirements.

An example of a capacitive accelerometer is shown in Fig. 7.19 [7.7]. Overload protection is automatically achieved by the capping

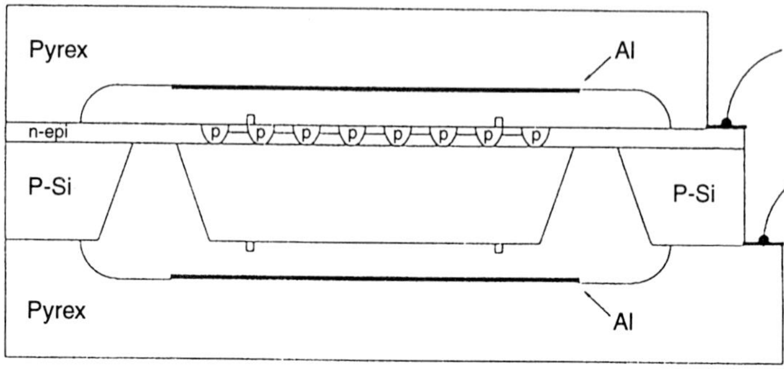

Figure 7.19. *Capacitive accelerometer [7.7].*

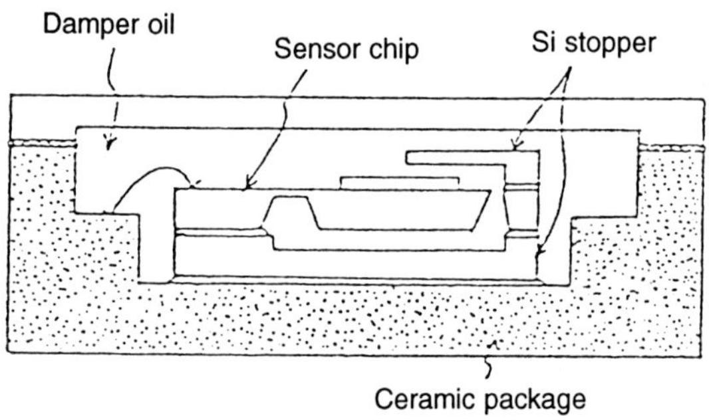

Figure 7.20. *Cross-section of piezoresistive sensor with extra overload stopper and damping oil [7.8].*

top and bottom since a shock will result in the mass coming against the upper or lower capping. Damping is achieved by air and is due to squeeze-film damping. Such a device has been developed at Delft University and is a fully integrated smart sensor system.

Piezoresistive-based accelerometers with on-chip electronics have been available for some years and involved the addition of a bulk

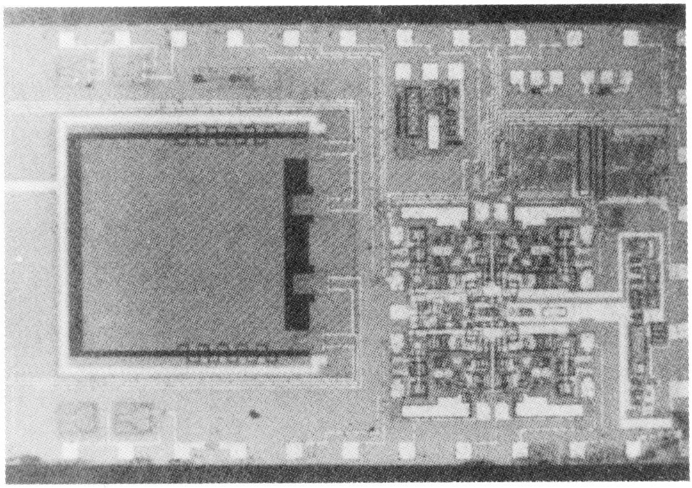

Figure 7.21. *Chip-photograph of a smart accelerometer [7.9].*

micromachining step after the electronics processing. Overload protection was achieved by bonding an additional silicon stopper to the sensing chip, as shown in Fig. 7.20 [7.8]. More recently there have been a range of structures proposed where the overload protection is incorporated into the sensor chip. In the case of this chip damping was achieved by using a damper oil. A later version of the piezoresistive-based sensor developed by Nissan is shown in Fig. 7.21 [7.9]. Compensation is provided for both the temperature effect on the resistors and any additional stress caused by the bimetal effect. The circuit for both half and full bridge configurations is shown in Fig. 7.22.

An increasingly important trend in the application of accelerometers is the reliability of the sensor. This is of particular importance for the air-bag application. The driver must have the assurance that when required, the air-bag will perform. The only way to ensure that this is the case is to test the device each time the automobile is started. For this reason there has been considerable interest in self-testing accelerometers and silicon is a material which lends itself well to such an application. This mechanism operates by the sensor generating a simulated acceleration and measuring the re-

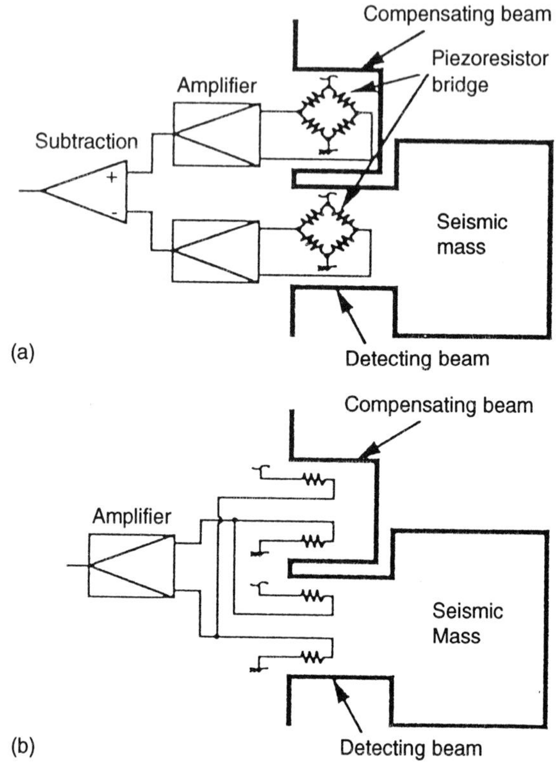

Figure 7.22. *Circuit configuration of the accelerometer shown in Fig. 7.20.*

sponse of the sensor. An example of a self-test mechanism is shown in Fig. 7.23 [7.10–7.11]. In this example the mesa is attracted to an electrode plate using electrostatic forces. This simulated acceleration is used as a test of the operation of the device. An alternative technique is to use thermal activation of the accelerometer, where the sensor structure is illustrated in Fig. 7.24 [7.12]. The test raises the temperature of the actuation beam by about 150°C, causing the beam to expand; this causes a downward movement of the mesa.

There has been considerable interest in silicon accelerometers

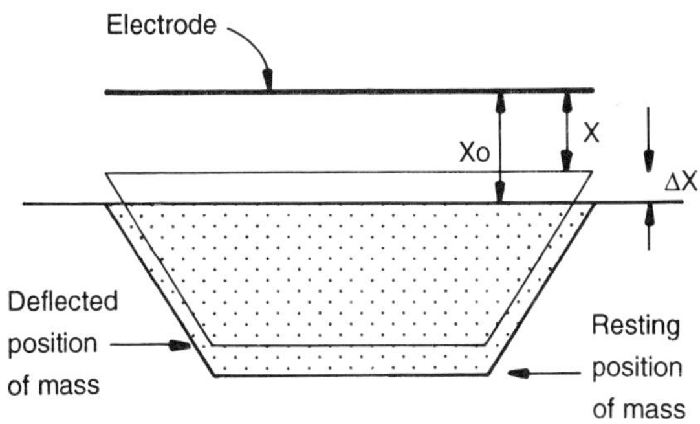

Figure 7.23. *Self-test mechanism for a silicon accelerometer based on electrostatic attraction [7.10–7.11].*

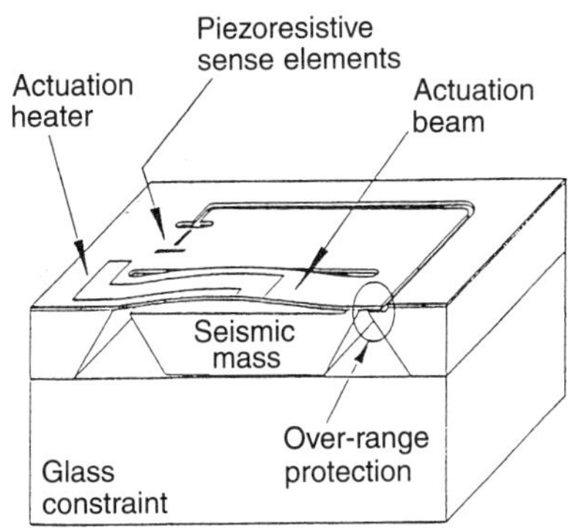

Figure 7.24. *Self-test mechanism for a silicon accelerometer based on thermal deflection [7.12].*

in recent years using, for example, the LIGA process [7.13], and looking at the performance of the device in more detail [7.14–7.15]. A quite different approach is a thermal-based accelerometer [7.16].

Accelerometers have a wide range of applications in automobiles all of which are suitable for silicon use. Silicon provides the possibility of on-chip compensation (cross-sensitivity and cross-axis

Table 7.10 Yaw rate sensor performance requirements for automotive applications

Range	± 60 deg s^{-1}
Overall accuracy	$\pm 2\%$
Resolution	± 0.1 deg s^{-1}
Bandwidth	0–20 Hz
Shock survivability	1 m drop on concrete
Temperature range	-40 to $+85°C$

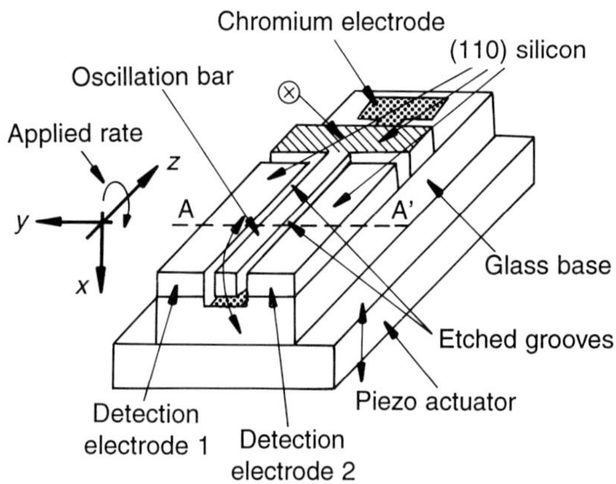

Figure 7.25. *Silicon micromachined gyroscope [7.18].*

sensitivity), improved S/N ratio (by measuring the acceleration and processing the signal on a single chip) and simple and compact self-test construction.

A further sensor of considerable interest is the yaw rate sensor, which detects the rotation of the automobile and may be applied to navigation systems. The required specifications of the sensor are given in Table 7.10.

As mentioned above, the gyroscopes used in the Cima automobile were optically based. There is considerable interest in the possibilities of fabricating gyroscopes in silicon, but until recently there have been few examples of working devices. Modern micro-

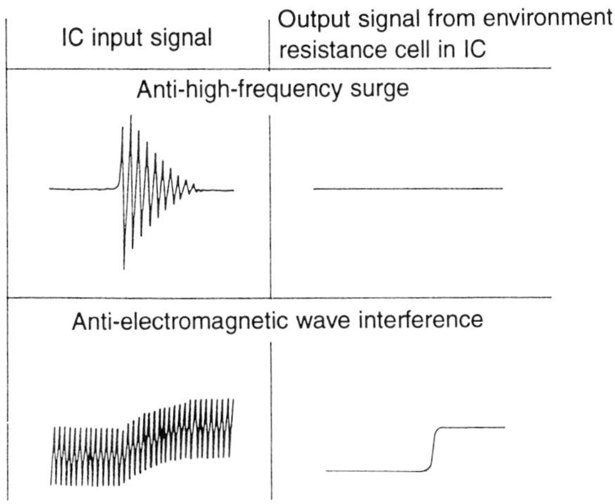

Figure 7.26. *Signals protected from environmental disturbances.*

machining techniques have opened the possibilities of fabricating such devices in silicon [7.17–7.18]. The structure of the micromachined angular rate sensor, based on an oscillating beam, is shown in Fig. 7.25 [7.18]. Due to the small capacitance which has to be measured, this is a prime candidate for a smart sensor to eliminate as much of the stray capacitance as possible and to amplify the signal as close to the capacitor as possible.

As mentioned above, the automobile is an extremely harsh environment with transient voltages and electromagnetic radiation. It is important to protect the electronics without too great an increase in the device size. At Nissan various techniques have been incorporated to the circuitry. One of these is the environment resistant cell which is incorporated into LSI circuits. The result is shown in Fig. 7.26. Since this makes the LSI itself more resistant to the harsh environment that thus the whole system is not greatly increased in size. Similar ideas have been incorporated into silicon power drivers.

7.6 Smart sensor possibilities for the future

Tables 7.11–7.13 list areas where smart sensor applications are of interest and where they are not. It can be seen that in many of the cases where smart sensor application is not attractive, the financial reasons are high development cost or cost competition. Cost competition means that the costs of sensors presently used are low enough and therefore there is no particular incentive to change. In some cases the technical reason for rejection is that the performance achieved with present-day devices is sufficient. In this case only a considerable reduction in price would convince managers to make changes.

The three most promising areas for smart sensors are:

1. Accelerometer for air-bag or chassis control system (suspension control, anti-lock brake system).
2. Yaw rate sensor for navigation system or chassis control system (hybrid navigation system).
3. Gear tooth rotation sensor for wheel speed, crank angle, gear rotation.

Accelerometers have been discussed for both ABS and air-bag systems. Recent years have shown a considerable increase in the number of papers describing silicon accelerometers using micromachining [7.7–7.16]. The three main requirements for automotive applications are:

1. Low cost — this can be met by the large-scale batch fabrication
2. High reliability — this is important for all automotive application, but in the case of the air-bag this is essential. Therefore, the development of a self-test mechanism for silicon sensors was extremely important. There have been several self-test systems presented in the literature [7.10-7.12], and some silicon accelerometers have already been applied in automobiles.
3. High S/N ratio (amplifier, temperature compensation circuits). This requirements leads to the smart sensor solution. A further important consideration is the effect of temperature on the mechanical properties of the device [7.16].

Yaw rate sensors for navigation or chassis control systems

Table 7.11 Future potential applications for smart sensors (electronically controlled engine)

System	Sensor	Suitable for smart sensor application	Technical considerations	Financial considerations
Electronically controlled engine	Air flow sensor	Yes	Relatively high resolution and wide range is required	Large number in use
	Air-to-fuel ratio sensor	No	High temperature (exhaust gas)	Development expensive
	Coolant temperature sensor	No	Low resolution, high pressure	Cost competition
	Cylinder pressure sensor	No	High temperature, high pressure	Development expensive
	Exhaust gas temperature sensor	No	High temperature, low resolution	Cost competition
	Throttle angle sensor	Yes	High resolution, contactless measurement	Large number in use
	Crank angle sensor	Yes	Digital output, contactless measurement	Large number in use

are finding increasing interest in the Japanese market. Examples include the hybrid navigation system and GPS (Global Positioning System) and the gyro. The requirements of these systems are:

1. Low cost — a few dollars
2. High S/N ratio (amplifier, signal processing circuits, temperature compensation circuits, etc.)
3. High reliability (self-check circuit)

Table 7.12 Future potential applications for smart sensors (electronic meter, navigation system and air-bag system)

System	Sensor	Suitable for smart sensor application	Technical considerations	Financial considerations
Electronic meter	Fuel level sensor	No	Present devices sufficient	Cost competition
	Washing fluid level sensor	No	Present devices sufficient	Cost competition
Navigation system	Vehicle direction sensor (yaw rate)	Yes	Easy to fit ECU; high resolution	Large number in use
Air-bag system	Acceleration sensor	Yes	Easy to fit ECU self-test	Large number in use

Recent advances in micromachining techniques have opened many new possibilities for integrating gyroscopes [7.17–7.18].

The third area of particular interest is the gear tooth rotation sensor for wheel speed, crank angle, gear rotation. ABS systems where the wheel rotation is measured will be increased in the future, and this will represent an extremely large market when it is introduced as standard. The requirements of the system are:

1. Low cost — a few dollars
2. Digital output signal corresponding to gear tooth
3. Simple interface (two wire interface)

In the continuing advancement of automobiles in Japan, several large projects have been started. These include:

- VICS — Vehicle Information and Communication System
- UTMS — Universal Traffic Management System
- ARTS — Advanced Road Transportation System
- ASV — Advanced Safety Vehicle

The last of these, ASV, is illustrated in Fig. 7.27.

In Europe, the US and Japan there are major projects being

Table 7.13 Future potential applications for smart sensors (electronic active suspension, ABS and TCS and air conditioning/air cleaning)

System	Sensor	Suitable for smart sensor application	Technical considerations	Financial considerations
Electronic active suspension	Acceleration sensor	Yes	High resolution, small size (easy to fit ECU)	Cheaper than nonsilicon alternatives
	Height and stroke sensor	Yes/No	High resolution, contactless meas. (antiabrasion)	
ABS and TCS	Wheel speed sensor	Yes	Digital O/P small size	Large number in use
Air conditioning; air cleaning	Room temperature sensor	No	Present devices sufficient	Cost competition
	Outside temperature sensor	No	Present devices sufficient	Cost competition
	Sunlight sensor	Yes	Higher performance might be required	Large number in use
	Smoke sensor	Yes	Various kinds of smell should be analyzed	
	Air mix door position sensor	No	Low resolution	Cost competition

undertaken to develop total smart systems where the vehicle can communicate with the road system. Nissan, like all the Japanese motor manufacturers have been developing 'smart' control systems for automobiles. Recent developments have included fuzzy control for adjusting cruise speed. These systems have been included into an automatic vehicle which combines communication systems with a range of sensors to establish conditions of the surroundings. The vehicle is able to drive automatically to a pre-determined destination.

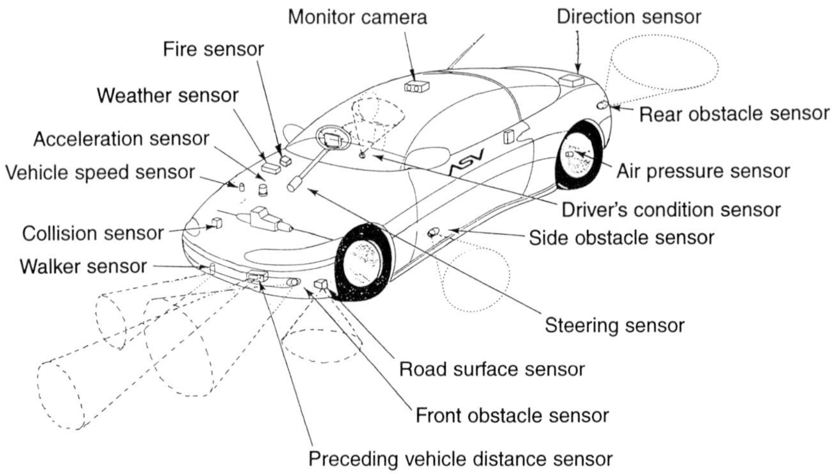

Figure 7.27. *Sensors for an advanced safety vehicle.*

7.7 Conclusions

It has been shown that there are considerable technical and financial demands on the designer who wishes to introduce smart sensor systems into automobiles. There are several applications where the technical problems can be overcome, but the cost of development or installation will, for the present, prevent application. It is furthermore extremely important to show that these devices are at least as reliable, and preferably more reliable, than the devices available on the market. Projects presently underway in Japan show an increase in the application of silicon sensors and where it is possible to take the next step to smart silicon sensors.

7.8 References

[7.1] C.O. Nwagboso ed., *Automotive Sensory Systems*, Chapman and Hall, London, UK.

[7.2] J. Krammer, Integration of smart sensors and actuators into automotive systems, Presented at Microsensors and Microsystems for Automotive Applications, NEXUS workshop, Frankfurt, Germany, 15 September 1994.

[7.3] M.H. Westbrook and J.D. Turner, *Automotive Sensors*, IOP Publishing, Bristol, UK.

[7.4] H. Muro, Optical angle detection sensor IC with high resolution, *Proc. 6th Sensor Symp.*, Tsukuba, Japan, 1986, pp. 295–297.

[7.5] P.J. French, H. Muro, Y. Hiramoto and W.Lian, An angle-of-rotation sensor using flipflop sensor techniques, *Sensors and Actuators*, A21-23 (1990) 414–419.

[7.6] S. Prosser, Microsystems for chassis management systems, Presented at Microsensors and Microsystems for Automotive Applications, NEXUS workshop, Frankfurt, Germany, 15 September 1994.

[7.7] R.P. van Kampen, M.J. Vellekoop, P.M. Sarro and R.F. Wolffenbuttel, Application of electrostatic feedback to critical damping of an integrated silicon capacitive accelerometer, *Sensors and Actuators*, A43 (1994) 100–106.

[7.8] M. Nakamura, K. Murakami, H. Nojiri and T. Tominaga, Novel electrochemical micro-machining and its application for semiconductor acceleration sensor IC, *Proceedings Transducers 87*, Tokyo, 3–5 June 1987, pp. 112–115.

[7.9] H. Muro, H. Kaneko, S. Kiyota and P.J. French, Stress analysis of SiO_2/Si bi-metal effect in silicon accelerometers and its compensation, *Sensors and Actuators*, A34 (1992) 43–49.

[7.10] H.V. Allen, S.C. Terry and D.W. de Bruin, Accelerometers systems with self-testable features, *Sensors and Actuators*, 20 (1989) 153–161.

[7.11] H.V. Allen, S.C. Terry and D.W. de Bruin, Accelerometers systems with built-in testing, *Sensors and Actuators*, A21–23 (1990) 381–386.

[7.12] F. Pourahmadi, L. Christel and K. Petersen, Silicon accelerometer with new thermal self-test mechanisms, *Technical Digest IEEE Solid State Sensors and Actuators*, Hilton Head, 22–25 June 1992, pp. 122–125.

[7.13] M. Strohramann, P. Bley, O. Fromhein and J. Mohr, Acceleration sensor with integrated compensation of temperature effects fabricated by the LIGA process, *Sensors and Actuators*, A41–42 (1994) 426–429.

[7.14] Y. de Coulon, T. Smith, J. Hermann, M. Chevroulet and F. Rudolf, Design and test of a precision servo accelerometer with digital output, *Proceeding Transducers'93*, Yokohama, Japan, pp. 832–835.

[7.15] K. Yamada and T. Kuriyama, A novel degree of freedom separation technique in a multi-axis accelerometer, *Sensors and Actuators*, A43 (1994) 120–127.

[7.16] R. Hiratsuka, D.C. van Duijn, T. Otaredian and P. de Vries, A novel accelerometer based on a silicon thermopile, *Proceeding Transducers 91*, San Francisco USA, 1991, pp. 420–423.

[7.17] J. Söderkvist, Micromachined gyroscopes, *Sensors and Actuators*, A43 (1994) 65–71.

[7.18] K. Maenaka and T. Shiozawa, A sudy of silicon rate sensors using anisotropic etching technology, *Sensors and Actuators*, A43 (1994) 72–77.

8
CONCLUSIONS

The integrated silicon smart sensor has been defined as a device fabricated using microelectronic processing technology and extensions thereof, that allow the on-chip functional integration in silicon of all the functions of a data-acquisition unit; viz. the sensor, signal conditioning, A/D conversion and digital systems for interfacing a sensor bus. The processing used for the fabrication of the sensor and the steps used for the realization of circuits should not interfere (fabrication compatibility, chapter 1). Special sensor materials could be deposited to add transduction effects where silicon falls short as a sensor material (compatible materials, chapter 3).

Unfortunately, the problems and challenges associated with on-chip functional integration are too diverse to enable a conclusion based on general statements and universally applicable guidelines. Nevertheless the most severe restrictions and opportunities can be identified. This book has basically been about two constraints for the successful realization an of on-chip integrated silicon data-acquisition system; viz. the technological feasibility and its economic viability. The initially claimed advantage of silicon as a sensor material, i.e. the mere material compatibility with integrated silicon circuits, has been shown to be an insufficient prerequisite. Fabrication compatibility is required for merging the two. Adding sensor processing should not interfere with proper functioning of the microelectronic components.

Throughout the chapters, it has become evident that the technology is not a major limitation, as special compatible sensor processing can usually be added. Therefore, the economic limitation is often reached first and this limits the market share of the device. A number of reasons have been identified that have caused the limited impact of integrated silicon smart sensors. Basically, these may be classified according to four aspects; viz. (a) those due to the market, (b) those due to the device, (c) those due to the structure of the sensor industry and (d) those due to the application in which the sensor is to be implemented.

8.1 The market

The market for silicon sensors is in measurement and control systems, which are to be implemented in consumer products, professional equipment and by the utilities. The measurement and control systems market is, therefore, basically a subcontractor of the final system assembler. This market is highly diverse due to the wide variety of end-products and very conservative due to the emphasis on proven reliability often (indirectly) imposed by the final system-user. As a consequence high-volume production is rarely achieved and the costs to enter the market are very high.

As discussed in chapters 4 and 5, high-volume production is a requirement for actually exploiting the first promise of the silicon sensor, i.e. the low unit cost in mass production already achieved in integrated circuits. If these economies of scale would actually apply, silicon sensors could have enormous potential in automotive applications. However these chapters also indicate this is not the case due to the fact that, unlike integrated circuits, the costs of bonding, packaging and testing exceed those of wafer processing. Moreover, the automotive industry is very demanding both with respect to the technical specifications, such as operation under severe environmental conditions at high reliability, as well as to the economic performance in terms of cost competition with the already implemented device (chapter 7). Nevertheless, the silicon accelerometer is very promising in this application, although, mainly owing to the excellent mechanical stability rather than to the prospect of functional integration.

Integrated silicon smart sensors offer a huge potential for improvement of reliability due to the reduced number of components and wire bonds. Moreover, RFI is minimized due to the small length of the interconnect used. However, as it is an emerging technology, data on reliability is scarce.

As indicated in chapter 5, the learning curve for silicon sensors is very slow when compared to the microelectronics industry and the cumulative volume above which the unit price would drop dramatically has not really been reached. Moreover, the silicon sensor would have to compete with conventional and mature sensor technologies that are on the falling slope of their learning curve or have dropped to a minimum.

An additional characteristic of the market is the preference for a new sensor that provides the same type of behavior as the conventional sensor (the replacement market), so that the remaining system components need not be replaced and the operating procedures on the system need not be changed. This implies, for example, that the output current of a silicon PTAT thermal sensor needs to be modified to provide a temperature-dependent resistance with which the user of a platinum resistor is familiar. The silicon sensor, therefore, has to add performance or functionality to obtain a competitive price/performance ratio, which brings up the other advantage of silicon, viz. its fabrication compatibility with readout circuits. In a way the limited success of silicon sensors is, therefore, due to the fact that the basic sensing element has not yet been integrated with the microelectronic circuits.

Research on an integrated silicon smart sensor is usually driven by a technology push, whereas a market pull would be a more sound approach from the economic perspective. However, a large part of the silicon sensor industry is presently mainly concerned about acute practical problems, such as bonding glue dispenser uniformity and pick-and-place equipment for fragile micromachined devices, whereas the research on integrated silicon smart sensors is mainly performed at institutes and universities.

8.2 The device

The second aspect is the integrated silicon smart sensor itself. The purpose of the device is to improve user-friendliness and thus market acceptance by eliminating the sensor non-idealities. Moreover it is to be realized on the smallest material carrier; the chip. This objective comes from the rapid market acceptance of integrated circuits. There are, however, a few constraints that are summarized here.

It has to be emphasized that the boundary between hybrid silicon sensors and on-chip integrated sensors has become fuzzy because of wafer-to-wafer bonding and flip-chip bonding techniques that are widely in use. As indicated in chapter 4, the merit of on-chip functional integration has to come from the added value this approach provides in the intended application. A sensor is subject to cost competition, which means that it is rarely purchased because of the sophisticated technology used for its fabrication, but rather because of its superior price/performance ratio, flexibility or functionality. In applications where conventional devices perform relatively simple functions satisfactorily, the integrated silicon sensor does usually not provide sufficient added value to replace it. However, when a certain application becomes feasible due to the on-chip functional integration, then the added value is maximum. The smart sensor concept is, therefore, attractive in (a) devices that are vulnerable to parasitics and electric interference at the output terminals (such as capacitive sensors), whose disadvantageous effect can be minimized by using on-sensor-chip amplification and (b) array sensors that require on-sensor-chip multiplexing to limit the number of bonding wires. The potential of new sensor technologies can also be judged from the application-dependent added value. As shown in chapter 4, it is worth investigating SEG/ELO in microstrip radiation detectors, but not for applications as an etch-stop.

Unlike microelectronic circuits, wafer yield is not the dominating cost in integrated silicon smart sensor manufacturing; there are also costs for dicing, bonding, testing for electrical and non-electrical parameters and non-standard packaging. Dicing and bonding are a general problem in sensors, but especially in micromechanical sensors (chapter 2), because of the stresses imposed on the fragile micromachined structures.

Unlike microelectronic circuits, not only the electric parameters need to be specified, but also the non-electrical parameters need to be tested extensively in a sensor in order to characterize the sensor performance in terms of e.g. the sensitivity to the desired non-electrical signal and the parasitic non-electrical signal. Moreover, testing is cumbersome, as it is often only possible on completed sensors, as e.g. packaging-induced stress strongly affects the offset specification of a Hall plate. The latter problem adds significantly to the costs, as rejected devices have gone through the complete process.

The sensor packaging is very expensive, not only due to the limited volume production mentioned already, but also due to the fact that the sensor package should on the one hand (like in microelectronic circuits) provide protection against environmental hazards,whereas on the other hand (unlike circuits) the sensor requires by definition access to the non-electrical measurand. These two requirements demand a package that allows selective access. Sensor packaging has widely been recognized as the major obstacle in sensor commercialization, as it is often neglected in basic research and results in a long development period before the sensor can be marketed.

The above-mentioned factors apply to both the basic sensor element and only to a slightly greater extent to the integrated silicon smart sensor and can, therefore, hardly be used as arguments against on-chip functional integration in silicon.

The wafer yield has been found to be an unsuitable measure to judge the smart sensor concept in chapter 4. In integrated circuit manufacturing wafer processing dominates costs. Increasing die area results in a reducing number of operating devices per wafer (yield) and thus in increasing cost per integrated circuit. Adding a sensor to an integrated circuit also increases the total die area, but due to the dominating costs of testing and packaging, the total costs are hardly affected.

An enhanced level of on-chip functional integration increases the density and thus also complicates testing. However, specially integrated 'disposable circuits' that are only used for testing purposes (e.g. multiplexers that allow wafer testing of sequentially selected individual dies) facilitate testing and reduce costs. In this context it should be emphasized that any processing performed after dicing causes a dramatic increase in costs. The costs per wafer-processing step are split among all the dies, whereas die processing directly attributes to the costs of that particular device.

8.3 The industry

The third aspect relevant in integrated silicon smart sensors is the sensor industry in relation to the microelectronics industry. As mentioned in chapter 5, the learning curve for the typical silicon sensor is very slow compared to that of the microelectronic circuit, which makes the silicon sensor market less attractive. As indicated in chapter 5, the curve is probably slow due to the fact that expertise is concentrated in small companies.

Obviously, manufacturing an integrated silicon smart sensor can be approached from the integrated circuit manufacturer perspective by using compatible sensor processing to fabricate the sensing element or *vice versa*. The major problem for the sensor industry is that the smart sensor concept aims for improved market acceptance and thus for increased volume production, which makes the device interesting to the microelectronics industry. As indicated in chapter 4, the 'knowledge barrier' is sensor type-dependent and causes the sensor company that has found a niche to be less vulnerable. However, it should be noted that the niches that were originally present in the microelectronics industry disappeared as soon as the learning curve had reached the downward slope.

Another complication is due to the fact that the cross-over volume in the unit costs versus volume beyond which the unit costs of on-chip functional integration are less than those of hybrid devices takes place at a lower volume in the microelectronics industry compared to the sensor industry, because of available equipment.

The sensor industry has so far flourished by parasitizing on the available infrastructure, both in terms of the availability of high-purity substrates at low costs and the network of equipment suppliers, that is the result of the mere presence of a huge microelectronics industry. Wafer-to-wafer bonding makes it possible for a silicon sensor company that has found its niche to minimize vulnerability by cooperating with the microelectronics industry by using this as a subcontractor of wafers that contain the readout circuits and mate these wafers in house with the wafers containing the sensing elements using relatively simple equipment. Basically, such an approach would add the microelectronics industry itself to the available infrastructure. However, the limits of abuse of this infrastructure also have come into view, because the specifications on equipment as imposed by the sensor industry exceed those imposed by the microelectronics industry. This issue has become relevant with the need for e.g. double-sided alignment and pick-and-place equipment for fragile wafers. Obviously, the development costs for such equipment have to be furnished by the sensor industry, which increases the added value of the silicon sensor required for it to become competitive. It should be noted that there is much room for improvement in the sensor industry, as there has been relatively little work done so far on e.g. manufacturability.

8.4 The challenge

The final aspect is the application. In this aspect the claimed advantage of small dimensions is of importance. As outlined in chapter 6, the small dimensions are crucial in biomedical applications. The low unit costs in mass production do not apply due to the fact that this application does not involve a consumer market. The compatibility and packaging problems are aggravated as biocompatibility is also required.

As mentioned already, silicon sensors have good prospects in automotive applications, however not primarily due to the on-chip functional integration of readout, but rather due to the excellent mechanical properties and the accurate and reproducible micromachining techniques available.

An important aspect to consider when judging the merit of an integrated silicon smart sensor in an application is that it constitutes a complete data acquisition system. It is therefore justified to compare it with a complete data-acquisition module composed of a basic sensing element, analog chips and digital building blocks rather than with any analog or digital IC. The integrated silicon smart sensor replaces that complete module and it is therefore unfair to estimate the required price/performance with that of only the basic sensor or an analog IC.

Summarizing, it could be stated that the integrated silicon smart sensor is a building block with huge potential in many applications; however, it is not as universal a building block as an integrated gate or operational amplifier and its merit does not provide sufficient added value in as many applications as were perceived at the introduction of the concept. Sensor systems based on wafer-to-wafer bonding and multi-chip modules will prove to be more cost-effective solutions in many applications. Nevertheless, full on-chip integration remains the ultimate solution and is unsurpassed in e.g. medical applications. Extensive research on compatibility is still required, with an emphasis on the most promising applications.

The increasing number of sensors in consumer products justifies the major research effort on integrated silicon smart sensors.

GLOSSARY

a	Diameter of micromachined circular membrane
A_c	Integrated circuit die area
A_d	Die area
A_s	Sensor die area
A_w	Wafer area
ABS	Anti-lock Breaking System
A/D	Analog-to-Digital
ADC	Analog-to-Digital Convertor
ARTS	Advanced Road Transportation System
ASSP	Analog Sensor Signal Processor
ASV	Advanced Safety Vehicle
b	Width
B	Magnetic field
BHF	Buffered Hydrofluoric acid
BiCMOS	Bipolar CMOS
BJT	Bipolar Junction Transistor
BPSG	Boro-Phospho-Silicate Glass
C_{cb}	Collector-base depletion capacitance
C_{dg}	Drain-gate capacitance per unit of width of the gate
C_{ox}	Gate oxide capacitance per unit area
CCD	Charge-Collecting Device
CMOS	Complementary Metal-Oxide Semiconductor

D	Diffusion coefficient
D_E	Diffusion coefficient in the emitter layer
D_n	Diffusion coefficient of n-type carriers
D_p	Diffusion coefficient of p-type carriers
D_0	Average defect density defined as the number of defects per wafer, N_d, over the wafer area, $A_w : D = N_d/A_w$
D_{0c}	Average defect density in the integrated circuit chip
D_{0s}	Average defect density in the sensor chip
$D_{p,ox}$	Diffusion coefficient of phosphorus in silicon dioxide
$D_{p,Si}$	Diffusion coefficient of phosphorus in silicon
DAC	Digital-to-Analog Convertor
DCS	Dichlorosilane
DI	De-ionised water
DMSO	Dimethylysulfoxide
DSP	Digital Signal Processor
DSSP	Digital Sensor Signal Processor
E	Electric field
E_{crit}	Critical electric field at breakdown
E_g	Electic field on the gate
ECM	Engine Control Module
ECU	Electronic Control Unit
E-D NMOS	Enhancement-Depletion N-channel MOS
EDP	Ethylenediamine-pyrocathecol-water
EEPROM	Electrically Erasable Programmable Read Only Memory
ELO	Epitaxial Lateral Overgrowth
EPROM	Electrically Programmable Read Only Memory
FSO	Full Scale Output
G	Geometric scaling factor
GPS	Global Positioning System
h	Height
HAc	Acetic acid
HF	Hydrofluoric acid

i_b	AC base current
i_c	AC collector current
I	DC current
I_c	DC collector current
$I_\mathrm{c,k}$	Knee current
I_e	DC emitter current
I_f	Forward current
I_s	Saturation current
IPA	Isopropyl alcohol
ISFET	Ion-Sensitive Field-Effect Transistor
JFET	Junction Field-Effect Transistor
k	Bolzmann's constant $= 1.380658 \times 10^{-23}$ J K^{-1}
KOH	Potassium hydroxide
l	Length
L	Diffusion length
L_B	Diffusion length in the transistor base
L_n	Diffusion length of the minority charge carrier in the base
L_p	Diffusion length of the minority charge carrier in the emitter
LOCOS	LOCal Oxidation of Silicon
LPCVD	Low-Pressure Chemical Vapor Deposition
LSB	Least Significant Bit
LSI	Large-Scale Integration
LTO	Low-Temperature pyrolytic Oxide
M	Mass
MAP	Manifold Air Pressure
MEMS	Micro-Electro-Mechanical System
MOS	Metal Oxide Semiconductor
n_i	Intrinsic free charge concentration in silicon
n_poly	Free charge concentration in polysilicon
N_B	Doping concentration of the base

N_{drain}	Doping concentration of the drain
N_{E}	Doping concentration of the emitter
N_{epi}	Doping concentration of the epilayer
N_{H}	Layer doping concentration in a Hall plate
N_{poly}	Doping concentration of the polylayer
N_{source}	Doping concentration of the source
N_{ss}	Idem
N_{sub}	Doping concentration of the substrate
N_{well}	Doping concentration of the well-layer
NRE	Non-Recurring Engineering
OCP	Open Circuit Potential
P	Power
P_{diss}	Power dissipation
PECVD	Plasma-Enhanced Chemical Vapor Deposition
PID	Proportional Integrating Differentiating controller
PP	Passivation Potential
PROM	Programmable Read Only Memory
PSG	Phospho-Silicate Glass
PTAT	Proportional To Absolute Temperature
PVDF	Polyvinylidene fluoride
q	Charge
Q_{b}	Majority carrier charge in the base per unit area
Q_{depl}	Charge of the depletion layer
Q_{ss}	Charge of the source
r_{H}	Hall constant
R_{th}	Thermal resistance
RAM	Random Access Memory
RFI	Radio-Frequency Interference
RIE	Reactive Ion Etching
ROI	Return On Investment
S	Sensitivity
SAW	Surface Acoustic Wave

SEG	Selective Epitaxial Growth
SEM	Scanning Electron Micrograph
SIMOX	Silicon Implanted with Oxygen
SOI	Silicon on Insulator
t_{epi}	Thickness of the epilayer
t_{H}	Thickness of a Hall plate
t_{ox}	Gate oxide width
t_{poly}	Thickness of a polysilicon layer
T	Absolute temperature (Kelvin)
TCS	Traction Control System
TEM	Transmission Electron Micrograph
TMAH	Tetramethylammonium hydroxide
UTMS	Universal Traffic Management System
V_{A}	Early voltage
$V_{\mathrm{br,npn}}$	Breakdown voltage of the npn transistor
$V_{\mathrm{br,pnp}}$	Breakdown voltage of the pnp transistor
$V_{\mathrm{cb,br}}$	Collector-base breakdown voltage
V_{dd}	Drain potential
V_{es}	Emitter-substrate voltage
V_{f}	Forward junction voltage
V_{gs}	Gate-source voltage
V_{i}	Intrinsic potential
V_{ss}	Source potential
V_{T}	Threshold voltage of a MOS transistor
V_{Tn}	Threshold voltage
V_{Tp}	Threshold voltage
VICS	Vehicle Information and Communication System
W	Width
W_{B}	Basewidth
x_{j}	Junction depth
Y	Wafer yield

Y_{hs}	Yield of a hybrid system
Y_{ss}	Yield of a single chip system
z	Vertical displacement
α	Clustering parameter
α_F	Common-base current gain
β_F	Common-emitter current gain
ε_{ox}	Oxide dielectric permittivity
ε_{Si}	Permittivity of silicon
λ	Photon wavelength
σ	Internal stress
τ_s	Storage time constant
φ_{depl}	Potential charge due to the charge stored in the depletion layer
φ_f	Electrostatic potential
$\varphi_{f,cb}$	Build-in potential
φ_{gs}	Difference in work function between the gate material and the silicon
φ_{ss}	Potential charge due to surface charge
ω_0	Resonance frequency

INDEX